jane grosslight

illustrations by jeffrey w. verheyen

effective use of daylight and electric lighting in residential and commercial spaces

light

A SPECTRUM BOOK

Prentice-Hall, Inc.
Englewood Cliffs, New Jersey 07632

Library of Congress Cataloging in Publication Data

Grosslight, Jane.
    Light, effective use of daylight and electric lighting
in residential and commercial spaces.

    "A Spectrum Book."
    Bibliography: p.
    Includes index.
    1. Dwellings—Lighting. 2. Commercial buildings—
Lighting. 3. Daylight. 4. Electric lighting. I. Title.
TH7975.D8G76 1984   729'.28   83–27082
ISBN 0–13–536300–4
ISBN 0–13–536292–X (pbk.)

A SPECTRUM BOOK

10 9 8 7 6 5 4 3 2

This book is available at a special discount
when ordered in bulk quantities. Contact
Prentice-Hall, Inc., General Publishing Division,
Special Sales, Englewood Cliffs, N.J. 07632.

Cover design by Hal Siegel
Book design and page layout by Maria Carella
Manufacturing buyer: Edward J. Ellis

ISBN 0-13-536300-4

ISBN 0-13-536292-X {PBK.}

Prentice-Hall International, Inc., *London*
Prentice-Hall of Australia Pty. Limited, *Sydney*
Prentice-Hall Canada Inc., *Toronto*
Prentice-Hall of India Private Limited, *New Delhi*
Prentice-Hall of Japan, Inc., *Tokyo*
Prentice-Hall of Southeast Asia Pte. Ltd., *Singapore*
Whitehall Books Limited, *Wellington, New Zealand*
Editora Prentice-Hall do Brasil Ltda., *Rio de Janeiro*

contents

foreword

Seldom does a "how to" book present its subject in such a light, lively manner as well as Jane Grosslight's book. I am fascinated by the wealth of usable ideas she has accumulated during her career as a lighting consultant.

You can do something about lighting problems and lighting opportunities if you read this book. By taking just one idea and putting it to work, you too can light up your life.

T. David McFarland, managing director
American Home Lighting Institute, Chicago

preface

Anyone who lives in an owned or rented space and anyone who works in an office or operates a business can use this book. It is intended for those who want to install the lighting themselves or hire someone to do it. It tells how to get the most usable light for the least energy. It gives *see-for-yourself* demonstrations to show lighting concepts and descriptions of lighting hardware. It tells how to use daylight effectively and how to use electric light efficiently.

Anyone who designs lighting—architects, lighting consultants, manufacturer's reps, interior designers, electrical engineers, lighting-showroom staff members, landscape architects, contractors, or electricians—can use this book. It gives *rules of thumb*, technical details, measurements in feet and meters, color values and their effect on light, and interior design tricks for augmenting or reducing daylight. It covers residential and small commercial space applications, combining information from architecture, landscape architecture, and interior design.

It took seven years to produce this book. It was created to be a comprehensive guide for lighting design by assembling scattered information and technical material and presenting them in an easy-to-understand manner. Particular attention was paid to include the many technological advances achieved by the lighting industry.

Special thanks to the many lighting professionals for their help and comments, especially Rita Harold, John Barton, Jim Jensen, Nancy Christensen, Noel Florence, Sam Zitter, Jim Yorgey, and Jim Nuckolls. I deeply appreciate the assistance I received from Florida State University's Center for Professional Development and from Lightolier and Lutron for being foresighted manufacturers and by providing me with grants to create the artwork.

I hope you enjoy reading and using this book for your own purposes; I enjoyed writing it.

light
your spaces

Do you need to light your spaces? Yes, you do. In residences, for instance, you need to light your spaces if you have a room that is dark and gloomy, if you are relocating or redecorating, if you want to make your living space more personal, if you prefer more colorful surroundings, or if you want more security. These and many other reasons should influence you to light your spaces, whether they are rented or owned.

In commercial spaces, you can improve your staff's productivity, stimulate sales, or create a pleasing merchandising environment with good lighting. Lighting has powerful effects for businesses, and you can use it to your best advantage.

Light is very important in any space. Colors and forms cannot be seen without it. All interior finishes and furniture are wasted without it, and functions cease. Remember the last time the lights went out? If you did not have a candle, you probably went out too, or went to bed. Most people take light for granted unless they are without it. They put up with the amount they receive from whatever fixtures deliver it. Unfortunately, lighting is rarely designed; it just happens. Fixtures attached to a structure are usually determined by the builder or the electrical subcontractor, who often decides on the basis of minimal necessity and minimal price. In residences, occupants bring in lamps chosen for their decorative qualities, not for their light-delivering capabilities. In the same way, occupants of commercial spaces bring in desktop lamps or showcase lighting and also try to make do. With this book, however, you can design good lighting for the effects you want.

Good lighting is like music; it gives pleasure and it gives cues, both physical and emotional. It gives a sense of enclosure, direction, and communication. Too much light is like noise; it detracts from the environment and irritates. Too little light is drab, gloomy, and dull, depriving us of color, glitter, and emphasis of highlight and shadow. In poor light, your space is always somber, like an overcast day.

You can light your space without lighting up your utility bill if you use daylight effectively and if you use electric lighting judiciously. Daylight is often not taken into consideration as a light source. It should be. It is free light, as long as it does not also bring in unwanted heat in hot climates.

Electric light can be used judiciously and can be designed efficiently. The lighting recommendations in this book are energy-efficient. The recommendations are tailored to many differences—personal, structural, and visual.

Personal differences include economics, age, life-style, and ownership. Economics dictates how much money is available to spend on lighting. But good lighting need not be expensive, just well planned. People of different ages

Do you have a room that is dark and gloomy?

Light your space without lighting up your electric bill.

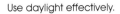

Use daylight effectively.

have different requirements for light. But, people of the same age have different lighting preferences. Likewise, life-style affects lighting choices. Ownership also affects lighting considerations; owners may build in lighting, while renters must not alter their structure and may want to remove the fixtures if they move. However, renters of commercial spaces often must leave lighting fixtures behind. The constraints of short-term owners and owners of mobile homes make their lighting solutions similar to those of renters. Likewise, owners planning to build a new structure have different needs than owners planning to remodel their lighting—identified in this book as relighting—which does not always require altering the structure.

Structural differences include characteristics of the space. The structure affects what kind of fixture can be used, how the light is received and reflected, and where the fixtures can be placed. These characteristics ultimately shape decisions about lighting design.

Finally, visual differences include characteristics of the activity that affect the amount of light needed. For example, critical tasks require more light than casual ones, and small details require more light than large ones. Lighting design must be tailored to these differences.

Contrary to what you might think, professional lighting design does not start with decisions about the fixture. The first decision is what is to be illuminated; the second is how it can be illuminated; and the third is which fixture is capable of doing it. This book provides information in written and pictorial form to help you make these decisions. Some illustrations are stylized, interior perspective drawings showing what lighting does. Some are section-cut drawings containing technical information. The fixtures described can be found in lighting consultants' or interior designers' catalogs and in lighting showrooms or department stores, depending upon the availability and degree of specialization of lighting in your area. About 200 manufacturers make lighting fixtures, and 3,000 lighting showrooms display them.

Everyone needs light, and everyone who buys lighting fixtures needs lighting information to make good decisions. This year alone, more than 100,000 housing units will have lighting installed before the structures are completed. Both owners and renters will occupy these new spaces. Occupants of spaces already built have opportunities to relight to suit their own style of living, to brighten their own life, and to increase the value of their home. In cities and small towns, spaces of 300 to 3,000 square feet (sq ft), or 28 to 279 square meters (sq m), are occupied by small businesses. Some are boutiques, beauty salons, or bakery shops; some offer insurance, income tax assistance, or investment advice. All have a product or a service to represent. All can improve their businesses with lighting, for very little money and a great deal of long-range payback.

This book gives many choices for lighting residential and small commercial spaces. It provides practical know-how, up-to-date information, and how-to directions. It defines terms as they appear in a sentence. Directions are detailed for do-it-yourself types. The same directions can be used by those that hire others to do the job. The strong-minded will be enthralled; the less resolute will be enticed.

This book dispels common myths about lighting and challenges the general acceptance of "making do." It provides rules of thumb to simplify technical lighting principles, points out differences that impinge upon lighting decisions, and supplies alternatives for changes—physical, mental, geographic. It describes aesthetics, compares sample electric costs, and describes alternatives for each lighting system. Terms are clarified within the sentence. You are shown how to get more light for less money. The scope of the book includes the wide range from necessities to luxuries. You are urged to identify your personal preferences, are taught how to use energy-efficient lighting, and are given both daylight and electrical lighting information for inside and outside. Use it for selected readings now and as a reference later.

how well are you treating yourself?

2

Most people think that the electricity they use for lighting is the main cause of their high electric bills. It is not. The main cause is heating anything electrically, such as heating water, heating the house, heating food, drying clothes, or blow-drying and curling hair. Electricity use for lighting, even considering inflation and fuel adjustments, is low. For example, lighting a 100-watt bulb for 1 hour at 10¢ per kilowatt-hour costs only 1¢. On the other hand, heating water electrically is eight times as expensive as electricity for lighting. Therefore, do not try to cut down by simply turning off the lights. You need light for your comfort and protection.

- Do not walk up or down stairs in the dark. (The medical bill for an injury would be far more expensive than the fraction of a cent it costs for electricity.)
- Do not cut, chop, sew, or saw in dim light. (Injuries are painful as well as expensive.)
- Do not watch television in the dark. (Illuminate the room with a light from elsewhere to balance the light from the screen.)
- Do not read for a long time in a room illuminated by only one light, even if it is directly on your book. (The contrast of the dark room and the bright book strains the eyes.)

Likewise, manage energy consumption in residences wisely by following these suggestions:

- Turn off incandescent lights—the bulbs in table lamps—when leaving a room, even if just for a minute. (This saves some electricity.)

- Turn off fluorescent lights—the lights usually found in office ceiling fixtures—when leaving a room for more than 5 minutes. (Turning them off too frequently, however, shortens the life of the tubes.)
- Use fluorescent instead of incandescent light whenever possible. (Fluorescents utilize electricity more efficiently and give more light for the watts consumed.)
- Use fewer appliances that heat and substitute more energy-efficient appliances for the uses required.

In small commercial spaces, manage energy consumption by following these suggestions:

- Do the whole space to the same level.
- Use the most energy-efficient lighting source possible—high-pressure sodium for small commercial spaces combined with daylight, metal-halide for a high-ceilinged store where shopping is done quickly, fluorescent adapters or bulbs for table lamps in intimate shops. (These sources actually can enhance merchandise.)
- Light only things that are important to be seen.

In general, be frugal by turning off lights that are not needed, by turning on enough when needed, and by using daylight whenever possible. Light is needed. Our living spaces need to be comfortable, pleasing, reassuring, interesting, and functional. A well-lighted space helps us do what we want and makes us feel good while we do it. Does your space measure up? If not, you are not treating yourself very well.

Reprinted by permission of Tribune Company Syndicate, Inc.

how much light do you need?

Our eyes are exceedingly adjustable; they adjust to moonlight and to bright sunlight. Rarely do our eyes call attention to the fact that the lighting is bad. However, bad lighting affects us all. The effects may be subtle and can occur almost unnoticeably (a sense of fatigue) or later (a headache), or the effects can be serious (eyestrain). The brain has difficulty judging the actual amount of light present, and does so by comparing the amount now to the amount received a few minutes ago. If it was bright and now is not, it is judged to be dark. If, on the other hand, it was dark and still is dark, the eyes adapt, and we feel that the light is sufficient unless it fails to satisfy our expectations or activity needs.

Amount Required for Activities

The Illuminating Engineering Society (IES) has developed a range of illumination amounts needed for everyday activities; it is measured in footcandles (the standard measure of the amount of light falling on a surface). The range is based on the age of your eyes, the reflectance value (amount of light able to be reflected), and demands of speed and accuracy of the activity.

The reflectance value of the activity is determined by colors of surfaces. Dark colors reflect little light; pale colors reflect more light, but never 100 percent.

Generally in residences, casual activities—everyday living and moving about the space—require up to 20 footcandles of light. Moderate activities—grooming, reading, and preparing food—require up to 50 footcandles. Extended activities—hobby work, household accounts, and prolonged reading—require up to 150 footcandles. Difficult activities—sewing, color matching, and other close work—require up to 200 footcandles. If, for instance, the eyes are older, the reflectance value is low, and speed is required, the highest number of footcandles need to be used. If the eyes are young, the reflectance value is high, and accuracy is unimportant, a lower number of footcandles is acceptable.

Generally in commercial spaces, circulation can require up to 30 footcandles. Merchandising requires up to 100 footcandles. Feature displays require up to 500 footcandles. Specific visual activities, such as drafting and reading fifth carbon copies, require from 200 to 2,000 footcandles. (For greater detail, consult the material in the *IES Lighting Handbook 1981 Application Volume* cited in the Bibliography.)

Measuring Light

How will you know if you have achieved the recommended amount of light? Your eyes cannot tell you, but you can measure light with a photographic light meter, the one you use for your camera. If the meter indicates amounts of light in footcandles, read how much is reflected from a white card with the light turned on at

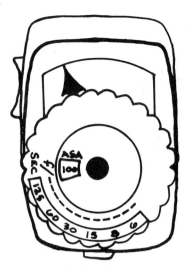

Photo meter.

night. (Do not point the meter toward the light.) If the meter indicates the amount as an exposure setting for a camera, translate it into footcandles as follows:

1. Set the film speed at ASA 100.
2. Aim the meter at a white card (a file card) where the activity is to take place (at a tabletop, on your lap, and so on).
3. Read the shutter speed opposite the f-stop of 4 as a whole number; the number is the footcandles of light available. (Technically, the reading is really in footlamberts; see Chapter 21.)

For example, set the ASA at 100 and read opposite the f-stop of 4. If it indicates 1/30 second, read it as 30 footcandles; 1/50 is read as 50 footcandles. For writing a thank-you note, 30 footcandles should probably be sufficient for most eyes in most environments. However, your circumstances may be different. Get enough light to satisfy your needs.

amount of light for a space

In Residences

Sometimes, activities take up a whole space or change from place to place within the space. For instance, family rooms can have dancing, dice games, reading, and relaxing. These activities occur on the floor, at a table, on the sofa, or all around the room. Use uniform lighting—even illumination throughout the space—and illuminate the activities wherever they are, and re-

duce the contrast between the activity and the other surfaces. One way to light uniformly is to match the size of the space with the total wattage used in the space.

- For small spaces (less than 150 sq ft, or 14 sq m), use approximately 200 incandescent watts or 100 fluorescent watts.
- For average spaces (up to 250 sq ft, or 23 sq m), use around 300 incandescent watts or 150 fluorescent watts.

- For large spaces (more than 250 sq ft, or 23 sq m), use 1 incandescent watt or ½ fluorescent watt per square foot of space.

Sometimes a residential space contains one activity at a fixed location. Light that activity; then light the rest of the space as needed. At other times a space contains many activities at fixed locations. First, light each activity, and use the lights one at a time or all together; the rest of the space may not need additional light.

Spaces used for relaxation should be lighted softly to permit circulation and recognition of the interior features and objects of interest, such as art or architecture. Spaces used for vigorous activity should be lighted brightly, revealing everything possible. All outdoor spaces should be lighted for safe passage, security, and orientation. Any living space, indoors or out, can be pleasurable and comfortable if it is lighted in the amount needed, at the location needed, and with a compatible color of light.

In Commercial Spaces

Commercial spaces must be lighted to accommodate activities, and should be changed when the activities change. For instance, if the circulation path is used minimally, light it minimally at 10 footcandles. If it gets more traffic, light it to 30 footcandles. When displaying merchandise, if the goods are highly recognizable and the time needed for the customer to evaluate them is short, light them brightly. Goods not easily recognizable and requiring assistance for a sale need only minimal light.

Some commercial spaces, such as restaurants, require atmosphere. Lighting goes a long way toward creating the necessary atmosphere for a restaurant. In any commercial space, the activity must be lighted first, and then the rest of the space should be lighted, for whatever reason.

light sources

For the purposes of this book, electric light sources can be divided into four categories:

incandescent (the kind in table lamps) ,

fluorescent (tubes in office ceiling fixtures), , high-intensity discharge (streetlights), and cold cathode (neon signs). The category of high-intensity includes mercury (the bluish streetlights), metal-halide (the cool-looking industrial lights), and high-pressure sodium (the yellow-looking highway lights).

Color of Light

All light sources have a characteristic color produced by the materials used to create the light. Within each category there are many types of bulbs, and their colors vary. The deluxe versions of all bulbs have the best color.

Incandescent light is a warm color, somewhat yellowish. Fluorescent light is cool, somewhat bluish, if cool-white types are used. However, other fluorescent sources are available that produce a warmer color that is compatible with incandescent light—warm white, warmwhite deluxe, and prime color fluorescent. Whenever incandescent and fluorescent sources

are combined in interior spaces, use only fluorescent sources that produce warm-color light. Many people think that fluorescent light belongs only in offices, but it can be very pleasing in interior living spaces if installed aesthetically and compatibly with incandescent-color light.

Deluxe color-corrected mercury is the warmest color of the mercury sources, which are normally considered bluish. Metal-halide usually is icy blue, but color-corrected versions are warmer. Indoors in commercial spaces, when metal-halide lights are combined with incandescent or fluorescent lights, both of these sources enhance the space and the objects in it. They cost less to operate than other sources, and can have a reasonable payback for their more expensive fixture and bulb cost. When they are turned off, most of them require a 10-minute cool-down period before they can be turned on again. These sources are particularly suitable as indirect light or as direct light with high ceilings in commercial spaces. In residential spaces, metal-halide sources may become acceptable in lower wattages without the 10-minute cool-down period. Outdoors these sources combine well enough with incandescent floodlights.

High-pressure sodium produces golden

yellow light, like concentrated sunshine. The color-improved type—less yellow—has been used successfully indoors in commercial spaces where it has been diluted with daylight or softened as indirect light. Outdoors at night the golden glow is more apparent, but as more streetlights are changed to sodium, people will become accustomed to them.

Cold cathode, or so-called neon, has a color that is dependent upon the gas and the color of the glass tube. Cold cathode light can be white, red, blue, yellow, or almost any hue. It does not give off satisfactory light for detailed visual tasks, but it does give off sufficient light to attract attention, indoors or out.

Color Uses

The color of an object appears to change when illuminated by different light sources. How the color appears depends upon the colors in the light source and the object's true color. In indoor commercial spaces where objects are sold on the basis of color, the colors in the light source are especially important. Objects need to be displayed under the same source they will be used under. For example, in a frame-it-yourself shop with only fluorescent cool-white light sources, a mat board and frame would be matched to the colors in the picture. At home, however, it might not look the same under incandescent light. Frames and mats might be returned by unhappy customers.

Likewise, colors reflected from the mirrors in dressing rooms and at the counter must be flattering. People trying on clothes will buy only if they like what they see. Therefore, light at a mirror is very important in commercial spaces. Incandescent light is the most flattering to all skin tones. Of the fluorescent sources, prime color or warm-white deluxe are the most flattering. Even if the color is right, the light can still be unflattering if it is too harsh, creating shadows or glare. Reflected light in the right amount can be soft and flattering for mirrors. (See Chapters 5 and 9.)

lighting systems

Turn off any lights that you do not need, but do not skimp on having light sources available. Lighting fixtures do not need to be all on or all off. Several individually switched fixtures can be used at different times—for instance, a lighted valance at the window, a lamp on the desk, a downlight over a coffee table, and fluorescent lights in the bookshelves. Used together, these fixtures should provide balance.

Also, several fixtures can be wired together and turned on at the same time—a lighting system. A system in a commercial space could include lights along the back wall that are wired together to turn on at night for protection. In a residence, a system could be a row of wall-washers or a series of fluorescent fixtures in a cabinet to brighten a formerly dark corner. Several lighting systems are even more effective when used together.

In residential spaces a lighting system or single fixture should be the primary source of light. The light should permit people to move around. It could also brighten the space on cloudy days, especially if it is low-wattage fluorescent. Contrary to standard practice, the primary light does not have to come from a center ceiling fixture. It can come from several fixtures (a system) that bounce light from elsewhere (a wall washed with light).

The secondary source of light should be used for specific activities. The fixtures for the secondary source need to be located where the activities take place, such as under a cabinet illuminating a countertop.

The third source of light may accent art, objects, architecture, or other special features. A single fixture (a downlight) or a system of fixtures (several downlights) can be used. The primary, secondary, and tertiary sources can be used together and enhance one another by balancing, by making the space appear more comfortable, and by giving more options.

In commercial spaces, the primary system of light should illuminate areas of activities and places that the customers should see—cashier stations, merchandise areas, displays, and so on. The secondary system of light should illuminate the space itself. The third system should provide light for nighttime surveillance. It can be a primary or a secondary system during the day, but should be located on the back wall of the space.

constraints

Lighting design, whether for new construction or relighting, is constrained by the structure to which the fixtures will be attached.

Sloped Ceilings

A sloped ceiling presents a constraint because the light is aimed in the direction of the slope, not straight down. All fixtures mounted on the ceiling (surface or recessed) will be affected. Such fixtures must be adjustable to compensate and aim the light straight down. Several types of surface-mounted fixtures are adjustable—canopy-mounted adjustables, eyeballs, and track fixtures. Recessed adjustable fixtures include internally adjustables, eyeballs, and telescoping spotlights. Some manufacturers make a "sloped ceiling adapter" for their recessed fixtures. Specify the angle of the slope when ordering. Chain- and cord-hung fixtures (pendants and chandeliers) compensate for the slope by the pull of gravity. Stem-hung chandeliers can also do so if they are made to be adjustable.

A wall along a sloping ceiling cannot be lighted uniformly because the wall is taller at one end than at the other. However, the ceiling can slope away from the wall without affecting the uniformity of wall lighting because the wall is always the same height.

Mobile and Manufactured Homes

Relighting mobile homes and manufactured housing also presents constraints in terms of wires and minimal number of baseboard receptacles (outlets for electricity). The construction of manufactured homes makes running wires in the walls difficult, and baseboard receptacles are usually far apart. Install wires in the walls from the outside by removing sections of the exterior finish material, and have an electrician put in new electrical wires. To reach existing baseboard receptacles with cords and plugs, carefully conceal and protect the electric wires by hiding them around windows and along baseboards. Never put wires under carpets or behind fabric unless they are especially made for such an installation and are heavily insulated. Codes prohibit using standard wires in such places. Surface-mounted standard wires can be covered with metal raceway channels and painted if necessary. Changing the wiring in the ceiling is easy because the ceiling panels are usually removable. To install wires or change ceiling fixtures, remove the ceiling panels from the inside by taking down the metal strips and taking out the staples. Then, have an electrician move the fixture to another place by connecting it to the old junction box with wiring, or install a

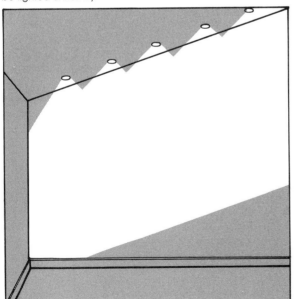

A wall with a sloping ceiling cannot be lighted uniformly.

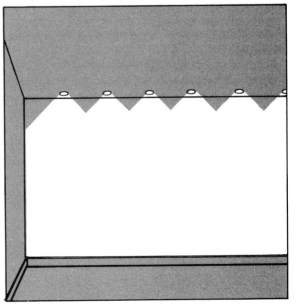

A ceiling sloping away from the wall does not affect the lighting.

new junction box. Finally, replace the ceiling panels with new panels from the manufacturer. Make sure the new panels match the old ones, or paint all of them the same color.

Concrete and Steel Construction

Relighting spaces built of concrete and steel is also constraining. Such structures do not accept recessed fixtures and do not permit rewiring in the walls, ceiling, or floor. They can be relighted by connecting fixtures to an existing electric outlet; by attaching wiring to the interior surface of the walls, ceiling, or floor and covering it with metal raceway channels; or by using track lighting with a cord and plug—provided it is acceptable to local electrical codes. (Tracks with cords and plugs are not allowed in some areas, especially in commercial buildings.)

Wooden Frame Construction

Relighting structures built conventionally (with wooden two by fours and wooden ceiling joists) is easy. Wires can be run in the walls, in the ceiling, and under the floor—especially if there is a crawl space in an attic or basement. Many electricians are skilled at wiring in seemingly impossible places in conventional construction. Be aware that electricians are trained to install fixtures, but are not trained to design lighting. You or a consultant must do it.

lighting consultants

Lighting design is a combination of science and art. It is not simple. Most lighting you see around you is poorly designed. It has not been purposefully designed to put light skillfully into a space. To imitate what you see most often is to repeat poor lighting practice. Lighting consultants can provide custom-design lighting. If you wish to use their highly skilled services, you

TASKS AND RESPONSIBILITIES IN LIGHTING DESIGN

	CLIENT	CONSULTANT	JOINT
Preliminary discussion			X
Contract			X
Design analysis		X	
Design proposal		X	
Conference on lighting design proposal			X
Selection of design	X		
Preliminary estimates		X	
Conference on estimates			X
Revision of design and estimates, if necessary		X	
Selection of electrician	X or	X	
Conference with electrician			X
Final lighting plan and fixture schedule		X	
Acceptance of final plan	X		
Awarding electrical contract	X		
Ordering lighting fixtures	X or	X	
Scheduling installation	X or		X
Inspecting and approving installation	X or	X	
Installation revision, if necessary			X
Lamping and focusing fixtures		X	
Evaluation of installation			X
Final evaluation	X		

Adapted from <u>How to Market Professional Design Services</u> by Gerre Jones, copyright © 1973, McGraw-Hill Book Co., New York.

have every right to ask several questions. Ask to see lighting jobs—either in photos or in person—that the consultant has designed. Ask for business references. Ask for a meeting of minds with the consultant you are considering, agreeing on what the two of you expect to do. Decide whether or not you want to have a large part of the decision making. If you do, you should have a good description of what you want and what you like. The more information you give a good consultant, the better he or she is able to serve you. (A client who is secretive or holds back probably will not be satisfied.) On the other hand, you might not know what you want. Tell your consultant. Consultants have many excellent design ideas, and you can choose what you like. The illustrations in this book will help you visualize how a design will look and may inspire some of your own ideas.

People hire consultants to light their residential and commercial spaces for many reasons. Some try to impress their friends or customers. Some try to get the right amount of light because they have eye problems. Some are not certain what they are trying to do because they have too many constraints impeding their decisions. Consultants can resolve constraints and describe impacts of lighting for easy decision making. Some clients do not want to make decisions. Consultants can suggest ways of lighting that would please anyone.

After choosing a consultant, you need to check to make sure that the cost and method of payment for the services are specified and to state clearly your financial limits for the lighting job. Then you can move ahead with confidence. However, you might want to interview a few consultants before you choose one. A wisely made choice is not often a source of regret. When the choice is made, the design process is a sequence of tasks for which the client, the consultant, or both are responsible.

doing it yourself

Many people enjoy learning and doing for themselves. It gives them a challenge and a sense of satisfaction. This book permits—in fact, encourages—doing it yourself. Contrary to what you might think, lighting is rich with choices.

**See for Yourself: How Would You
Light This Room?**

1. Look at the illustration of the bedroom and identify by name the lighting you would choose for the space (with no constraints of money or style).

2. Draw the lighting fixture where you would want it, or write the word for the type of fixture where you would place it.

3. How many choices did you identify? (If you drew the sun outside the window, you get an A for this lesson. Indeed, daylight is a valuable way to light any space.)

4. In addition, you can install electric lights in four places: at the head of the bed, at the dressing table, at the chaise lounge, and on the outside patio.

5. In most instances, people will think that a pair of lamps beside the mirror and a floor lamp beside

the chaise are the only choices of lighting for a bedroom, but at the least four different ways are available to light each place. Therefore, there are 16 combinations, not including the number of fixture choices (4 places times 4 ways).

6. Four ways to light four locations in this bedroom are:

At the head of the bed
   wall-washer fixtures
   wall lamp for reading
   cornice lights behind the drapery canopy
   fluorescent strips in the bookshelf niche

At the dresser
   recessed incandescent downlights in the soffit
   recessed fluorescent lights in the soffit
   a pair of wall lamps on either side of the mirror
   surface-mounted incandescent lights beside the mirror

At the chaise lounge
   recessed downlight
   pendant
   downlight in a chandelier
   floor lamp

Outside on the patio
   wall lamp on the brick wall to match the wall lamps beside the mirror
   uplight behind the large, potted evergreen shrub
   floodlight at the roofline of the patio

recessed downlights along the roof overhang over the glass window wall.

Why would you choose to light four places? Light at the bed, among other things, allows you to spread out your checkbook and bank statements or to read a novel and slip off to sleep. Light at the dresser allows for combing hair or matching clothes. It also enlivens interior colors (flowers and pictures) and creates highlights and shadows (on the antique mirror frame). At the chaise, light allows for watching television, sinking back and reading a magazine, or undressing for bed in soft background light. Outside, light on the patio can reveal the texture of the brick and provide an assurance at night that all is well.

Most often, bedroom lighting is a center ceiling light pumping 150 watts and causing shadows and glare, and a dresser or bedside lamp of 100 watts. However, their total (250 watts) could be spread around the room in different light sources and used more effectively only when needed.

These examples suit different needs. The figure below has enough light at the mirror for inspection but not for applying makeup. Figure center opposite has only enough light at the bed for short-term reading before falling asleep. Your lighting design needs to be tailored to your life-style and your needs.

At the bed—one 30-watt and two 15-watt fluorescent tubes in the canopy, equalling 69 fluorescent watts (including 15 percent more electricity for the fluorescent ballast).

At the dresser—two 15-watt bulbs in each wall lamp, equalling 60 watts.

At the chaise—six 6-watt incandescent bulbs in a chandelier and a 50-watt reflector downlight, equalling 86 watts.

The total inside is 215 watts.
(Outside on the patio wall, a lantern.)

At the bed—two 15-watt flourescent tubes in the bookshelf niche, equalling 35 watts.

At the dresser—three 50-watt downlights in the soffit, equalling 150 watts.

At the chaise—one 75-watt reflector in a downlight.

The total is 260 watts.
(Outside, a floodlight.)

At the bed—one 12-volt, 12-watt reading light (including 20 percent more electricity for the transformer).

At the dresser—two 40-watt fluorescent tubes recessed and ballast.

At the chaise—a 150-watt incandescent floor lamp.

The total is 256 watts.
(Outside, an uplight.)

At the bed—two 30-watt wall-washers.

At the dresser—ten 12-watt incandescent bulbs beside the mirror.

At the chaise—one 75-watt bulb in a pendant.

The total is 255 watts.
(Outside, a floodlight.)

chandeliers and other visible fixtures

3

Do you have a chandelier hanging over your dining table? Do you think that the room is therefore lighted? Most of it is not. Unfortunately, most dining room chandeliers are equipped to overpower and therefore produce poor-quality light. Usually they are equipped with over-bright bulbs. Sometimes they are wired through a dimmer, but when more light is required chandeliers are turned on fully. At that moment the bright, glaring light comes from the center of the room and tries to light everything. Anyone turning their back to the chandelier creates his or her own shadow. Anyone trying to look across the room is doing so through a blaze of light. A dimmer can subdue the blaze, but shadows and darkness are created instead. Shades on the chandelier can subdue the glare, but they also subdue the only source of light. One source is not enough. Chandeliers must be balanced by other sources to be comfortable and flattering.

Chandeliers and other visible fixtures—pendants and wall fixtures—are like pieces of jewelry, attractive but not necessarily functional. Even though they may be expensive, they are not intended to light a whole space, unless the space is small or used for a short time. Like any jewelry, they are flattering accessories. They are highly visible when lighted, so they should not be glaring. They should supply only soft and glittering light.

Adding other light sources in the same room as a chandelier will not necessarily increase the utility bill. First, all of the sources do not have to be used at the same time. Second, the total wattage now used in the visible fixture can be divided and used in several places around the room. The new total wattage may actually be less than the previous wattage. For example, if a chandelier has five light bulbs of 60 watts each, the total wattage will be 300, and the light will be harsh. Create a new total by:

- Reducing the wattage in the chandelier to 15 for each bulb, for a total of 75.
- Adding three 40-watt fluorescent fixtures behind a valance board, putting light up and down, over the draperies along a 12-ft (36.5-m) window wall, for a total of 120 watts.
- Adding two 30-watt reflector downlights over the buffet on the opposite wall, for a total of 60.

The new total is 255 watts, 45 watts less than the old total. Electricity is saved each time all the fixtures are on and the light is soft, sparkling, and well balanced. Often, the fixture can be used alone or in some combination. Seven possible combinations are:

- The chandelier alone, for setting the table (75 watts).
- The chandelier and the valance light, when looking for the serving platter (195 watts).

A chandelier is like a piece of jewelry.

- The valance light, the downlights, and the chandelier, for a gala dinner party (255 watts).
- The valance light and downlights alone, when dusting the furniture and drapes (180 watts).
- The valance light alone, for brightening up the space on a dull day (120 watts).
- The downlights and the chandelier, for a cozy dinner for two (135 watts).
- The downlights alone, for a nightlight when no one is in the space but you do not want a dark room (60 watts).

If there is a chandelier only, the space can be lighted only one way—by having the chandelier on or off. Why not save electricity and gain more light by using a combination of light sources? These can include wall fixtures, wall-washers, accent lights, downlights, cove lights, and lamps. Not all are on display in lighting showrooms. Most showrooms are a sea of visible fixtures, but other kinds of lighting are also available though not as often displayed. Check the index of this book for the types of light sources available, and ask lighting consultants about them. If you have or want visible fixtures, they can be beautiful, flattering, and energy-efficient jewelry. Learn how and where to use them.

Most visible fixtures need to be electrified.

They look best with the least hardware showing, hooked up directly into an electric junction box. This connection requires prewiring or new wiring. Otherwise, a cord and plug or electric track components are required. Owners, with the help of a good electrician, whether they are building or relighting (installing new lighting), can usually connect directly to the electricity. Renters usually cannot. They must rely on connections with a cord and plug.

Most visible fixtures require incandescent light bulbs. Such bulbs are available in a wide range of wattages, often specified by the manufacturer. However, the specified wattage is usually the highest and not necessarily the best to use. To create soft, flattering light, equip your fixture according to the following rules of thumb, based on whether or not the bulb can be seen.

Rules of Thumb for Incandescent Light Bulb Wattage

- For visible light bulbs, use up to 15 watts.
- For obscured but somewhat visible bulbs, use up to 25 watts.
- For hidden bulbs, use 40 watts maximum or the manufacturer's specifications.

Try the lowest wattages first; often the lowest one provides sufficient light. Your visible fixture should enhance the space, not overpower it.

A fluorescent screw-base bulb can substitute for an incandescent bulb in a fixture that hides the bulb, as long as the required amount of illumination is provided. Some pendants can accept an adapted-circle fluorescent tube or a screw-base fluorescent bulb, if it can be well hidden. Use any size that fits and gives you as much light as you want. Some visible fixtures are made specifically for fluorescent sources; use the one specified by the manufacturer.

Visible fixtures have decorative and human impact. They must relate to the scale of the room—to its length, width, and height. In general, avoid large fixtures in small spaces and small fixtures in large spaces. The decorative impacts of visible fixtures are the mass of the fixture, the intended impact, the proportions, and the harmony of the whole space. The human impacts are the visual comfort, the amount of light needed, and the unthreatening position of the fixture. Consider these impacts carefully and you will never regret where you have hung your lighting jewel.

chandeliers

Where to Use a Chandelier

Historically, chandeliers were hung in the center of a room, well above people's heads, to light as much of the room as possible. Chandeliers can still be used in the center of a room in period style and in some contemporary rooms for special effects. In addition, chandeliers can be used above all sorts of tables—dining tables, end tables, coffee tables, corner tables, and game tables. Likewise, a chandelier can be used above a desk, a bar, or any piece of furniture that is not a seat. A chandelier above a seat, hung at any height, can be threatening, making people uneasy. Linear chandeliers are called library or billiard lights. These chandeliers accentuate the geometry of any rectangular table. Over desks and library tables, they provide light to the right as well as to the left. Over billiard tables and dining tables, they illuminate the activity for the length of the table.

A chandelier can be positioned off-center above a buffet, dresser, or vanity top. In fact, small dining rooms appear larger when a chandelier is moved from the center of the table and placed off-center above a buffet against the wall. Small bathrooms are enhanced with a small chandelier to the left or right of a sink in front of a large mirror, reflecting the sparkle.

A contemporary ceiling fixture becomes a chandelier when centered over an eating table. Do not neglect these fixtures as a possibility for an effective piece of lighting jewelry.

Finally, a chandelier hung so that it can be seen through a window creates a pleasing architectural surprise, both inside and out. When the glimpse of it is momentary and later revealed fully, the visual excitement is heightened. Chandeliers hung in unusual places create a tasteful visual surprise. The possibilities of places for chandeliers are almost endless and probably limited only by one's pocketbook.

Linear chandeliers are called library or billiard lights.

A small chandelier in front of a large mirror enhances a small bathroom.

A contemporary ceiling fixture gives a chandelierlike effect.

A chandelier seen through a window creates a pleasing architectural surprise.

How to Choose a Chandelier

A chandelier can be of the same style as the furnishing in a room, thereby carrying out the harmony. For example, a wrought-iron chandelier is congruent with Early American, rustic, or contemporary furnishings. An Oriental faux-bamboo chandelier blends with Queen Anne, Georgian, or neoclassical furnishings. The possibilities of harmony are numerous. A chandelier can also be of a different style, thereby carrying out a counterpoint to the style of the room. Antique classic furniture can be combined with a gleaming chrome fixture, contemporary in style. Smooth Eero Saarinen furniture can be combined with a chandelier dripping with crystal. Create harmony or counterpoint; mix or match.

Size of a Chandelier

Hung in a room and not over a table, a chandelier must relate to the size and scale of the space. Rooms with high ceilings create a large volume and accept larger chandeliers than smaller rooms. High ceilings are enhanced with two-tiered chandeliers, which would be unsuitable in average-height spaces.

Rule of Thumb for a Chandelier Not Hung Over a Table

The chandelier can be in inches what the diagonal of the room is in feet. (For example, if a room is 16 ft on the diagonal, the chandelier could be approximately 16 in. in diameter.)

In the metric system the rule is that the chandelier's diameter can be in centimeters what the diagonal of the room is in meters divided by 0.12. (For example, a room 16 feet on the diagonal is 4.88 meters; 4.88 divided by 0.12 is 41, and the chandelier would be in proportion around 41 centimeters.)

Two-tired chandeliers enhance high-ceilinged spaces.

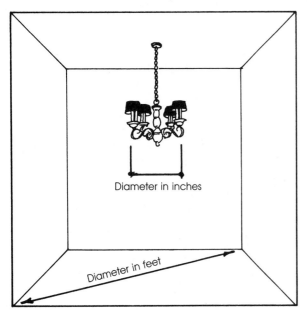

A chandelier can measure across in inches what the diagonal of the room measures in feet.

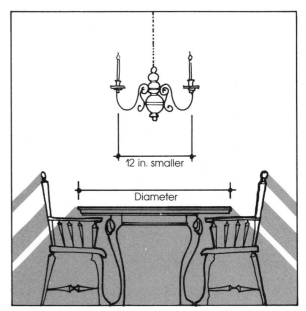

A chandelier can be 12 in. smaller than the diameter of the table.

This rule about size is merely a guide and can be modified for decorative reasons. If the chandelier is open and delicate, it appears to be smaller and a larger size might be suitable. If the chandelier is bulky and solid, it appears larger and acts as a visual barrier, so a smaller version might be more suitable. In general, big chandeliers need big rooms and small chandeliers need small rooms. The results can be breathtaking. A chandelier hung over a table must be small enough to allow people to get up and down without bumping into it.

Rule of Thumb for a Chandelier Hung Over a Table

The chandelier should be 12 in. (31 cm) smaller in diameter than the smallest table dimension.

At game tables, where the action is more vigorous, a chandelier should be even smaller. At rectangular tables a linear fixture (library or billiard chandelier) should not exceed two-thirds of the length of the table.

How to Hang a Chandelier

A chandelier hangs from the ceiling by one of three devices: a chain and cord, a cord only, or a stem. Chains and cords can be easily shortened or lengthened. Stems can be shortened, but cannot be lengthened. (Sometimes longer stems can be ordered from the manufacturer.) A canopy cover obscures the place where the hanging device and the ceiling meet. A chandelier is hooked up to the electricity in most cases through an electrical junction box, as required by many building codes. However, a junction box is not always located in the best place on the ceiling. Do not despair. Several alternatives are possible.

New ceiling junction box. Install a new ceiling junction box if electrical wires are accessible, that is, if wires can be pulled from elsewhere inside the ceiling. Access is gained through an attic or by pulling wires between the ceiling joists. If access is available, this is the slickest

Reprinted by permission of Oldden.

method. Owners who are building or relighting will want to use this method, even though it usually requires an electrician. An electrician's fee is not prohibitive when compared to the investment in the chandelier itself.

Another ceiling junction box. If another ceiling junction box is available elsewhere on the ceiling, attach the cord there and bring it over to the chandelier by one of the following five methods:

1. Weaving the chandelier cord through a chain and holding it by hooks from the other ceiling junction box.
2. Putting the chandelier cord into a canopy and extending the cord over to the junction box.
3. Knotting and hooking the chandelier cord over to the other junction box.
4. Hiding the wire from the junction box in a raceway channel (a U-shaped metal covering).
5. Connecting the chandelier to a track adapter and fastening it into an electric track wired at the junction box.

All of these connections are highly visible because hardware is put on the ceiling. Of the five, the electric track offers the opportunity to change the position of the chandelier, which is useful to some but wasted by others. The purchaser, of course, pays for this adjustability. However, as the popularity of track lighting has increased, the cost has come down. If it is desired, most chandeliers can be adapted to electric tracks. The limitations are the weight of the chandelier itself and the diameter of the cord or thread size of the stem, which connect to the track adapter. Check the manufacturers' catalogs to coordinate the chandelier's components with the electric track requirements before you purchase; they must be compatible.

Owners who cannot have a new junction box installed and renters who want to move their chandelier will want to use the electric track method of installation. (Consult Chapter 7 about the track clip-adapter method of chandelier installation before making a choice.)

Existing baseboard receptacle. A chandelier can be connected by a chain, cord, and hooks across the ceiling and down the wall to an existing baseboard receptacle. This method is often called swag. Swag is the last type of installation I would suggest, because it detracts from the aesthetics of the fixture; too much hardware shows. However, both owners who have purchased living units in structures that are not suitable for rewiring and renters who want to invest minimally in chandelier installation will

A chandelier should be 2 ft. 6 in. above the table.

want to use this method. An owner planning to build should never use it.

Once the electrical connection is made, the next decision is how high to hang the chandelier.

Rules of Thumb for Hanging a Chandelier

For center-of-the-room installation in average-height rooms, hang the chandelier close to the ceiling. In high-volume spaces, hang the chandelier lower so that it is still in view.

Above an eating table, the lowest part of the fixture should be no less than 2 ft 6 in. (0.76 meters) above the tabletop.

Off-centered over desks, sideboards, bars, and stairwells, the chandelier should be pretested to make sure that sufficient headroom is provided for people to move about and to use the space and furniture.

A head and a chandelier do not mix—one gets bruised, the other gets broken. Likewise, a chandelier must never be hung in a position that threatens a person—over his or her head, for instance. Human comfort is a lighting goal.

A Downlight in a Chandelier

When it is over tables or table-height surfaces, a chandelier becomes more than just a pretty jewel if it contains a downlight. A downlight provides direct, bright light onto the surface below. It can be a working light for short-term activity, such as writing a letter or reading the mail. In addition, the direct light will produce sparkle on tabletop objects: glasses, dishes, and silverware. If the chandelier is to be used over an eat-

Chandelier light.

Downlight and chandelier.

Downlight.

ing table, gain double sparkle—the sparkle of the chandelier itself and the sparkle of the downlight on the objects below.

In small rooms a downlight can take the place of other light sources. But it does not spread light broadly. However, if the table is the only furniture in the space, a downlight is especially effective.

Sometimes manufacturers specify a standard A bulb and sometimes a reflector bulb for a downlight. The latter packs a greater punch. Frequently, a 75- to 150-watt reflector is suggested, but 25 to 50 watts are often as effective and conserve more electricity. (Manufacturers indicate the maximum, not the minimum.) Consequently, try lower wattages first.

Do not use a downlight over a glasstop table. The table acts like a mirror. It reflects light from the downlight bulb. It glares at you wherever you are—sitting or standing—and is impossible to get away from. Therefore, in cases such as this choose a chandelier without a downlight.

In addition, chandeliers with or without downlights can create reflectances on any large glass surface, whether it is a mirror or an undraped window at night. A chandelier reflected in a mirror or in the glass of an undraped window can look doubly good, provided that the chandelier is not overpowering and is producing soft glimmering light. Brilliance should come from the decorative effect of the chandelier, not the wattage of the bulbs.

wall fixtures

Wall fixtures illuminate from the wall. They come in numerous styles:

- Sconce (half of a chandelier).
- Track fixtures on an electric track.
- Wall-hung lantern.
- Bar-shaped fixture with bulbs behind.
- Accent lights, pointing in one direction.

Where to Use Wall Fixtures

Use wall fixtures in a dining room above a buffet, on a large mirror, up the steps, gracing an arched doorway, beside the bed, or in other hard-to-light places. Furniture is not needed below them.

Use wall fixtures to illuminate semipublic business spaces where a less officelike appearance is desirable. Restaurants, shops, restrooms, waiting rooms, and professional offices can appear warmer and more inviting with appropriately styled wall-hung fixtures.

Wall fixtures can be used alone or in pairs. They can match a chandelier or another fixture in the room. They can be centered, off-centered, or used in pairs. However they are used, they can accessorize a space by themselves.

How to Choose a Wall Fixture

Sometimes you choose a wall fixture to enhance the style of the room or structure. If so, always choose a fixture that matches another visible fixture. For example, a pair of outdoor wall lanterns hung beside a Federal mirror above a dressing table match a lantern outside on a balcony. If there is no other fixture to match, choose one in a different style to enhance the interior design. It can be successful if carefully planned.

Some wall fixtures are adjustable. Some

Match a wall fixture with a chandelier.

A pair of outdoor wall lanterns hung beside a mirror look effective with a matching lantern on the outside balcony.

are stationary. Both provide light in difficult places. Use adjustable wall fixtures to light a desk, a bed, or an easy chair where there is no space for a table lamp. A track fixture and track canopy can function together as an adjustable wall fixture.

Some wall fixtures allow the bare bulb to be seen and should provide only a soft amount of light. Some fixtures direct the light above and below. Some point the light in one direction to accent objects. They are difficult to use successfully and need other light sources for balance and softening.

Sometimes a wall fixture can create a sense of scale. The size of the fixture must be compatible with the wall's height, the furniture, and the volume of the space. To determine the size, test it.

See for Yourself: Is the Wall Fixture the Right Size?
1. Build a rough model with cardboard and masking tape as close as possible to the shape and size of your wall fixture. (Shoe boxes are often the size of a wall fixture.)
2. Hold the model up to the wall.
3. You can not only judge the size but also decide how high to hang it.

How to Hang a Wall Fixture

There are two ways to connect a wall fixture to the electricity. Connect the wires to the electricity in a junction box behind the fixture in the wall. This installation is called wall mount or outlet-box mount. Owners who are building can obtain a junction box by having the wires installed originally, or owners who are relighting can pull the wires from a baseboard receptacle below the desired location of the fixture. Skilled electricians can pull wires from seemingly impossible places.

Using the wall mount or outlet-box mount to connect a wall fixture is the most aesthetic method because the cord does not hang down. The cord is objectionable to many people. Over and over again, the readers of my newspaper column on lighting expressed their displeasure at seeing cords.

If there is no junction box, plug the electric cord into a baseboard receptacle. This installation method is called pin-up, or cord and plug. When purchasing a fixture, specify which type of installation is to be used.

Renters must use the cord and plug method. Sometimes owners choose to do so. Be sure that the electric wire is long enough to reach the nearest baseboard receptacle. Otherwise, have the pin-up rewired. Some have hollow tubing or a chain to obscure the cord. The tubing comes in 1-, 2-, and 3-ft lengths (0.3, 0.5, 0.8 m). Check the manufacturer's catalog to see if such accessories are available for the fixture chosen.

Rule of Thumb for Hanging Wall Fixtures
All wall fixtures are hung at or near eye level.

Wall fixtures are meant to be seen. Hang them higher only if they infringe on the space needed to move around; for example, beside steps.

Uplight and Downlight in Wall Fixtures

Wall fixtures can be uplights and/or downlights. Be careful that no one is blinded by the glare or dazzled by the light, especially on the stairs, under balconies, in upper bunks, and in other up and down places.

See for Yourself: Is There a Glare?
1. Hold a flashlight where the wall light will be.
2. Have someone check the glare above or below.

Alternative Energy Sources for Wall Fixtures

Frequently, readers of my newspaper column would ask about battery-operated wall fixtures for their living spaces. They are available for utility spaces where electrical connections are difficult. They are, however, very utilitarian looking. For more formal spaces, wall fixtures using other energy sources are better.

Manufacturers make candle- and kerosene-fueled wall fixtures. Some are elegantly styled. The light produced by these fixtures is pleasing, comfortable, and never glaring. But it is produced by fire. Consequently, the fixtures require monitoring as well as cleaning. Notwithstanding these inconveniences, give these fixtures more than just a passing thought in your deliberations. Candle or kerosene light creates a festive mood.

pendants

Pendants are ceiling-hung fixtures similar to chandeliers but usually without branches. In some cases pendants spread the light. In other cases they push light down like a downlight. Transparent or translucent pendants spread light; opaque pendants push it down. If light needs to be distributed broadly in the space, choose a transparent or translucent pendant. If a worklight or a concentrated pool of light is wanted, choose an opaque pendant. These often create cozy or intimate light.

Hanging devices for pendants are like the ones for chandeliers—cords and chains, cords, or stems. Some cords have pulleys that make the pendant adjustable up and down. (The British aptly call pulleys rise-and-fall mechanisms.) A pulley is especially good for those who cannot or should not climb up to change a light bulb.

Where to Use a Pendant

A pendant can be used in the same places as a chandelier: centered above a table and off-center above other furniture except a seat. Over a seat, a pendant becomes a threat to one's head. Pendants appear natural over other furniture, and they also emphasize architecture without furniture, such as a dramatic open stairway, a high ceiling, or a massive fireplace. They call attention not by lighting the architecture but by being associated with it. They fit in small spaces, such as a narrow window. They can light an area that needs bright, direct worklight. Pendants can be clustered from a common canopy or hung in rows independently. They can be a repeatable design element.

Opaque pendants can create an intimate mood if hung at eye level, casting their light down at an eating table, at each end of a sofa, beside a bed or lounge chair—anywhere intimacy is intended. In professional offices, pendants make the atmosphere less institutional. Like chandeliers, they can be either very expensive or reasonably priced. Good judgment and good taste will help you choose a reasonably priced pendant that looks expensive.

How to Choose a Pendant

Pendants come in various shapes—domes, drops, lanterns, shades, and globes. Some pen-

A pendant can provide a repeatable design element.

dants emulate the styles of yesteryear, and some are in styles not associated with any historical period. Mix or match them with your interior design. Beyond style, pendants can emphasize activities. For example, a stained-glass pendant containing the Queen of Hearts, Jack of Diamonds, King of Clubs, and Ace of Spades over a game table reinforces the intended activity. Use what is appropriate for the space.

A pendant must relate to the size and scale of a space. A high-ceiling space can accept a larger pendant than a low-ceiling space. A small space must not be overwhelmed with a big pendant without first carefully considering the decorative effects. This is not to say that a large pendant should never be used; it can be, if it is carefully chosen. For example, a large, paper-covered Japanese lantern is appropriate in some small spaces. But, before buying, make a rough model in the general shape of the pendant being considered, hold it up, and judge it in the space.

Pendant emphasizes an activity.

The rule of thumb for the size of the chandelier given earlier in this chapter is also a guide for a pendant, although opaque pendants need to be smaller, usually by about 20 percent. If it is undersized, however, a pendant looks puny. Both too big or too small pendants are unsuitable. A lighting fixture should never stand out because of its size, whether it is large or small.

How to Hang a Pendant

Hang a pendant from the ceiling by hooking it up to the electricity through a ceiling junction box. If a ceiling installation is not possible, it can be hung from the wall as a swag. But a pendant looks best when connected directly through the ceiling. Another choice of hanging is to adapt it to an electric track. A track can reach from an existing ceiling junction box located in an undesirable location. A pendant on a track can be movable.

When hanging a pendant over a table, follow the rules of thumb for hanging a chandelier given earlier in this chapter, or follow these rules:

Rules of Thumb for Hanging a Pendant
- Alongside a chair or sofa, the bottom of the pendant should be about 42 to 45 in. (a little more than 1 m) from the floor.
- Beside a bed, the pendant should be 24 in. (61 cm) from the top of the mattress.
- Above a desk, the pendant should be about 20 in. (51 cm) from the surface.
- Not over furniture, the pendant should be hung at eye level.

Light and Color from Pendants

Pendants hung at eye level are governed by the rules of thumb for incandescent light-bulb wattage: Visible light bulbs must be 15 watts or less; somewhat obscured bulbs can be up to 25 watts; and hidden bulbs can be as high as 40 watts unless otherwise specified by the manufacturer. If a reflector bulb is specified for your pendant, use it. A reflector puts out a greater punch of light from a smaller wattage. Even if the manufacturer says the highest reflector wattage usable is 75, try 50 or even 30 watts. Lower wattage might produce enough light, and it will save electricity. Pendants hung above eye level can have higher wattage light bulbs, but be careful not to overpower the space with glare. Higher wattages need to be softened with other light sources.

If you want to get as much light as possible, or you prefer an active mood in the space, choose a pendant that allows the light to go up as well as down. Light will be reflected from the ceiling. If you want to achieve a more passive mood, use a downlight pendant. Light will be more concentrated and can be subdued. Make sure it is not glaring. Some downlight pendants have scientifically designed louvers to cut off the view of the bulb and reduce the glare from the bottom of the fixture. They are very good for seating areas. Always choose a pendant with a louver or lens if it can be seen from below. If the pendant is viewed from above, choose one that blocks the light. Comfort is worth considering when choosing and hanging a pendant.

Translucent pendants made of colored materials—glass or plastic—can cast a color with the light. Be sure the color does not create unpleasant effects. In addition, carefully coordinate the color of a pendant with the background against which it will be viewed. To blend with the least contrast, a pendant should be the same color as the background. To create a pleasing contrast, pretest a pendant of contrasting color against its background—if not with the actual fixture, at least with a piece of similarly colored construction paper. In addition, make sure the pendant's color is repeated elsewhere in the space so that it is not an intruder. If it is carefully chosen, a pendant can be an accent, not an accident.

selecting the light-bulb style for visible fixtures

Many different types of bulbs are available for visible fixtures, particularly for chandeliers and wall fixtures. If they are well chosen, they can flatter the fixture and the space. The list of types is long, but in most places the supply is short. Supplies are often clustered in big cities at electrical supply and lighting showrooms. Small-town showrooms do not usually have a large selection, but other stores in your area might have or be willing to order these special bulbs. Try grocery stores, discount department stores, building supply stores, nationally advertised department stores with mail-order catalog sections, and even drugstores.

Do not spoil an attractive chandelier or chandelierlike wall fixture with an indifferent bulb. Choose one that enhances it. Test the bulb in the fixture. Try several styles. Ten different styles are manufactured:

Shape Descriptions	Manufacturer's Identification
standard household A	A
teardrop	B
cone (often used in nightlights)	C
bent candle	CA
flame	F
globe	G
straight sides	S
tubular	T
chimney	GT

Manufacturers identify bulbs by a letter and number coding. The numbers indicate in eighths of an inch what the diameter of the bulb is at its widest point. Therefore, an A-17 is 17 eighths of an inch, or 2 ⅛ inches (5.4 cm) across.

The teardrop, bent candle, and flame shapes (B, CA, and F) are the traditional shapes for bare-bulb chandeliers and wall fixtures. Some flame bulbs are manufactured to flicker like a candle flame. However, these usually look better in restaurants than in homes. The globe, teardrop, and cone shapes (G, B, and C) are good for stylized chandeliers. These non-flame shapes are different, not indifferent. They give an additional decorative look. In contemporary chandeliers with clear glass chimneys, a tubular-shape bulb elongates the light source and is attractive. In chandeliers that are copies of gas chandeliers, tubular bulbs covered with a cloth mantle imitate a gas flame. The chimney bulb combines a light bulb with a glass chimney, thereby eliminating the necessity of purchasing both. They do not fool anyone, however. The standard household A bulb must be used only in chandeliers designed to hide the bulb, such as chandeliers with shades, opal globes, or frosted chimneys. Never allow a standard A bulb to be seen.

Most bulbs are manufactured in either a candelabra (small-size) or standard (medium-size) base to fit the socket size of any chandelier, though some are not. Adapters can modify a standard socket to accept a candelabra size. Be cautious; sometimes the proportions look awkward. Check out an adapter and a bulb in your fixture first, unless the bulbs are unseen.

Some bulbs will not work in an upside down or horizontal position. Ask your supplier before buying or check the manufacturer's catalog.

The glass in bulbs can be clear or coated. Clear glass permits you to see the filament.

Coated glass diffuses the filament. Some clear bulbs have a tint—iridescent, amber, ruby, smoke, blue, and green. The tint adds a slight hue to the light. Coated bulbs are red, orange, and so on, but white is the most common. Frosted bulbs are not quite the same as coated; they have a hot spot, because they do not diffuse as well as white coated bulbs. (The new energy-saving standard A bulbs in higher wattages are coated, but the coating is semitransparent. Treat them as though they were coated bulbs; do not use them as clear bulbs.)

**Rule of Thumb for Choosing
a Clear or a Coated Bulb**
Choose a clear light bulb when the light bulb can be seen, and always choose a coated bulb when the light bulb cannot be seen.

This rule sounds backwards, but it is not. Fixtures with clear glass and no shades are designed to show off the lighted filament. Consequently, if the filament is obscured by a coated bulb, the aesthetics of the design are also destroyed. But do not create glare by using more than 15 watts in a clear bulb.

If the manufacturer indicates that a visible fixture, particularly a downlight pendant, is rated for both a standard A and a reflector bulb, choose a reflector bulb and lower the wattage. A lower wattage reflector bulb will produce as much light as a higher standard A bulb, but it will consume less electricity. However, the fixture must have a porcelain or ceramic socket; brass sockets get too hot for reflector bulbs. If you are in doubt about the type of socket that you have, ask your fixture supplier or electrician.

lamps to light up your life

4

Lamps can light up your life. In small commercial spaces, lamps are not often used; fluorescent fixtures are. Consequently, lamps are refreshing. In any space they lend a sense of scale and add warmth not possible with ceiling fixtures alone. In residences, on the other hand, lamps are used too often. Lamps have limitations. In many ways, you'll find that other types of lighting described in this book are better for residences.

limitations

Usually the light created by lamps is an accident of placement, not the intent of lighting design. A lamp is limited by the length of its cord and can be positioned only where feasible. Consequently, within a space, light is created in spots here and there. The spots are visible on the furniture, the floor, and the ceiling. The ceiling spot can reflect and spread the light, but it can also be bright and distracting. Spots of light on the furniture and on the floor may be useful, but often they are not.

Lamps are also limited by the characteristics of the shades. Translucent lampshades can deliver light through the shade, but they become bright. Opaque shades cast light down or up in a size dependent upon the width of the shade's opening and in an amount dependent upon the bulb's wattage. The light can be harsh.

Lamps are usually limited to being near or on furniture. Table lamps are restricted to table-height furniture. Floor lamps are ordinarily, but not always, restricted to a chair, table, or group of furniture. The result is that light is not always in the best place.

In addition, lamps cannot systematically light anything. Neither table nor floor lamps, for instance, can wash a wall with light from end to end. Washing a wall with light makes a space appear larger and feel more hospitable. It flatters the occupants, and enhances the furniture. Do you know a lamp that can do that?

Finally, lamps are limited because they are chosen for their looks, rather than for the kind of light they produce. Clearly, different lamps and different shades produce different kinds of light.

benefits

In spite of their limitations, lamps do have benefits. They are so easy to install—just plug them into a baseboard electrical receptacle. They are portable and have instantaneous installation.

Lamps are excellent supplements to fixed lighting—ceiling lights, wall lights, chandeliers, and all built-in fixtures. Also, for many spaces, lamps provide light where no other lighting solution is possible. They can light difficult places—a tiny area or a place where other lighting cannot be installed. For instance, an arc lamp or stretch floor lamp can put light down onto a group of furniture in a steel-and-concrete condominium where a ceiling fixture cannot be used, or it can reach over a sofa to illuminate reading material where the space is not sufficient for a table and a lamp.

Wall receptacle.

Lamps have another function; they give a sense of scale to the space. Most sizable rooms should have one lamp. Choose a lamp in scale with the room and the furniture—big rooms with big furniture require big lamps; small rooms with small furniture need small lamps.

lamp choices

If lamps are selected for their looks only, they are simply a decorative ornament and should not be expected to provide adequate light for the space they light. Like a chandelier, which is also "jewelry," decorative lamps must be backed up with other light sources. If, on the other hand, they are selected to provide light for a specific purpose, they can be functional as well as decorative. Always choose a lamp for the type of light desired—confined and directed or broadly spread. Lamps that confine the light are those with opaque shades, such as metal-shaded desk lamps or torchères. Lamps that broadly spread the light have translucent shades, such as table lamps with off-white silk shades.

Light confined.

Light diffused.

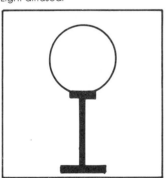

Direct and adjustable light.

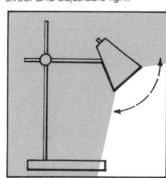

Transluscent shade.

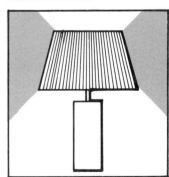

Lamp bases control the height from which the light will be distributed, and lampshades control the distribution of the light. Shades are either translucent (permitting the light to filter through) or opaque (not allowing the light to filter through). Never judge the color of a translucent shade until you see it lighted; it often changes color completely. Be aware of the distribution characteristics of shades.

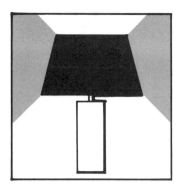

**Rule of Thumb for Distribution
of Light by Lamp Shades**
The wider the opening of the shade, the wider the light will be spread. The spread of light will be more noticeable with opaque shades than with translucent shades.

If a table beside a seat is to be illuminated, make sure that the shade is wide at the bottom. If a large area is to be illuminated by reflecting light from the ceiling, make sure that the shade is as wide at the top as at the bottom. To ensure that the shade functions as it should, follow the rules of thumb for the bottom of the shade; they vary with the position the shade will be in.

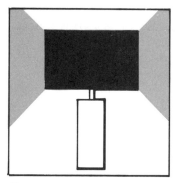

Shade wide at the top and bottom.

Rules of Thumb for the Bottom of the Shade
A lamp with a shade directly beside a person's head, such as a lamp on an end table, should have the bottom of the shade at eye level.
A lamp with a shade above eye level should be placed behind, at the right or left rear corner of the seating.
A lamp with a shade above a person's head, such as a floor lamp used as a downlight over furniture arranged in a group, should have a louvered or diffuser-covered opening to obscure the light bulb.

The inside surface of the shade affects the amount of light reflected. It should reflect as much as possible. Contrary to what you might think, metallic surfaces—silver and gold—reflect poorly and can be glaring. Overall, shades should be deep enough to hide the bulb, wide enough to spread the light, dense enough to obscure the lighted bulb, and lined with white to reflect the greatest amount of light.

Lamps themselves are highly visible and produce their light without any mystery about the source. Use the shade to protect the light bulb from being seen from the usual positions in the space—seated and standing. If a mirror is to be illuminated, make sure that the bulb does not glare into someone's eyes and that the light is transmitted to the face. Likewise, if a table lamp beside a chair is desired, do not select it until the height of the table and the chair are known.

light bulbs

The light source for lamps can be chosen from many wattages and several types, whether they be incandescent and fluorescent. They fit into the lamp sockets, which are either metal or ceramic (resembling white china).

Incandescent

Coated standard A bulbs are good for most lamps. They are manufactured in either single or triple (three-way) wattages. Use three-way bulbs if lamps are the only source for lighting. Determine where to place them for the best effects. Sometimes turn on two or three lamps at the lowest wattage; at other times turn on two at full and one at lower wattage. Control is at your fingertips. Make lamps work for you and save energy at the same time.

Sample Electric Cost

A three-way bulb used at 50 watts instead of 150 watts will save 100 watts or $1.20 per month at 10¢ per kilowatt-hour and 4 hours per night. However, the option for 150 is available when needed.

Some lamps, usually of better quality, indicate the maximum wattage usable. Do not exceed the indicated wattage. Some lamps do not indicate a maximum. Use a wattage suitable for your purpose. But do not overwatt a lamp so that it produces glaringly bright light; install other light sources. In addition, never install an incandescent reflector or a PAR bulb in a metal socket on a lamp; install it in a ceramic socket only.

Fluorescent

Fluorescent sources last longer, put out more light for the watts they burn, and are cooler than incandescent sources. Two types fit into lamps, one with an adapter and one without. The adapter type holds one or two circle fluorescent tubes and permits them to fit into an incandescent socket. Fluorescent circle tubes are available in several wattages—20, 22, and 40—and they produce different amounts of light. The 20- and 22-watt fluorescents put out the same amount of illumination as the 40-watt incandescent. They are suitable for hallways and small rooms where a low level of illumination is sufficient. They are not suitable for reading or other visual activities. The 40-watt fluorescent approximates the light from a 100-watt incandescent bulb. It can be used for a lamp beside a chair or mirror, if the room is not finished in deep colors and your eyes are good. All of them should be used in fixtures that are deep, wide, and opaque enough to hide them.

The fluorescent source without an adapter looks like an incandescent bulb; it has its own screw base. This source is small but it packs a big amount of light. The 18- and 25-watt size produce enough light for desktop tasks. Other compact fluorescent sources are made in 7, 9, 13, and 20 watts; they need a two-pin electrical connection and a starter in the lamp base. Lamp manufacturers are designing portable lamps for these small energy-efficient fluorescent sources. Use these compact fluorescent sources wherever possible.

amount of light

The amount of light seen in the space will depend upon the other light sources and the interior finishes. To observe how the light from more than one lamp blends with other lamps, test several lamps as described in Chapter 7. In general, strong directional light from an opaque shade in a dimly lit space will contrast with the surrounding darkness and be irritating. Likewise, an overly bright translucent shade against a dark wall causes too much contrast. Shades that blend with their background are usually considered more pleasing, such as pale-colored shades against pale walls and dark shades against dark walls. However, contrasting shades can be used if they are well planned.

Determine if your translucent lampshade is too bright when lighted. Follow the directions in the section on measuring light in Chapter 2, but put the meter in the middle of the shade to measure the footcandles.

Rules of Thumb for Shade Brightness

- For general illumination in a room with deep-colored walls or wood paneling, the shade should indicate no more than 50 footcandles on the meter.
- For general illumination with pale-colored walls, the shade should indicate no more than 150 footcandles.
- For illumination to perform specific visual activities, the shade should indicate between 250 and 400 footcandles.

---

designing with lamps

The best lamps have an undershade device to soften the light—a diffusing bowl, a refractor, or a disc. However, most lamps do not. Even though undershade devices improve the quality of the light, they are not always necessary if the light in the space is well distributed and blended. Design well-distributed lighting by utilizing the other lighting described in this book as well as lamps. Decide where light is needed for activities, what can be emphasized with the light, and how to balance it.

See for Yourself: Where, What, and How Much?

1. Determine where the electrical baseboard receptacles are and what furniture is appropriate to hold or be related to lamps.
2. Test various bulb wattages in lamps at these locations with both opaque and translucent shades.
3. Decide which positions, shade types, and bulb wattages distribute and blend the light satisfactorily.

Make light perform the way you want it to, and enhance your spaces—at work or at home—for many different moods, effects, and needs. In many ways the vocabulary used to describe how you want the light to look determines how you light it. If it is described simply as "light," you will not have any criteria to make decisions. Describe what you want with many adjectives and adverbs and the solutions will be easy.

Examples

- A dual-purpose light at the desk to evenly illuminate the reading material in various positions and to call attention to the colorful print on the wall above. (The solution is a neutral-colored translucent shade on an adjustable desk lamp.)
- A resplendent soft light at the coffee table, emphasizing the seating group and creating a subspace. (The solution is an arc floor lamp, a lighted coffee table, or, if possible, a metal pendant hanging from the ceiling, which would also create intimacy and emphasize the architecture.)

- A harmonizing, soothing light at the end of the seating group, bringing the dark corner back into view. (The solution is a table lamp with translucent shade compatible with the other lamp shades.)
- An incremental, confined light at the favorite reading chair. (The solution is an opaque floor lamp, positioned beside the chair with the bottom of the shade below eye level and equipped with a three-way bulb to provide low to high levels of illumination for quiet to busy activities.)

If lamps are the only light source for a space, observe the following rules of thumb.

Rules of Thumb for Lamps

- Never have only one lamp on in a space, unless the space is very small.
- Never read or perform other activities in a space with only one lamp lighted. (Additional light sources reduce glare, contrast, and shadows, permitting the eyes to function without excessive fatigue. All spaces should be without glare, but not all spaces should be without contrast and shadows. Contrast and shadows are desirable as long as they are not severe and harsh. Without them, living spaces would be dull and drab like a gloomy day. Overall, lamps should provide light that is adequate for seeing, diffused and distributed, glare-free, and free of severe contrast and harsh shadows.)
- Never position a lamp so that a shadow falls on the activity. (Heads, shoulders, and hands should not cause a shadow. The lamp should be placed so that it casts its light directly on the surface of the activity.)
- Have all lampshades at the same height, if possible. (This gives space a sense of unity. Other similar elements also do this, such as all the same metal and all the shades in the same unlit color.)
- For reading, the closer the light bulb is to the bottom of the shade, the more light will reach the reading material.

You need more light than you probably realize. The wide variety of lighting fixtures described in this book can provide that light, the easiest of which is lamps. However, use lamps with a lighting plan in mind.

An illuminated coffee table emphasizes a seating group.

lighting
a special
wall

5

Have you put money and effort into finishing one wall as a special wall? If so, you need to light it. Light permits the colors to be seen, emphasizes the design, and enhances textures. The light can either wash or graze the wall. Washing reveals the details and color of the carefully chosen wall covering, such as wallpaper, a mural, a wall graphic, wood paneling, fabric covering, or a contrasting paint color. Grazing reveals the intended texture, such as brick, stone, textured wallpaper, drapes, or other intentionally rough surfaces. Either type gives you the maximum benefit from the special wall, thereby justifying the expense. Further, light reflected from the wall adds to the other sources. Alone, it is sufficient for many occasions. Equally important, the reflected light flatters people in the space, and according to illumination research, people feel more positive about a space when the walls are lighted. Try it.

See for Yourself: Is Reflected Light Pleasing?

1. At night, in a room with a pale-colored wall, bring in one or two lamps with metal shades that throw the light in only one direction (for example, a reflector clamp-lamp or an auto trouble light).
2. Turn out all other lights.
3. Put the lamps close to the wall and direct the light toward it.
4. Observe the quality of the reflected light on people's faces and how the room looks.

Wall-washing is created by positioning the fixtures 1 ft (0.3 m) or more away from the wall and aiming them so that they cover it thoroughly and evenly. An owner can wash with a low-wattage, low-energy consuming system—either a system with low-wattage incandescent wall-washers or one with fluorescent cornice lights. A renter can use a portable wall-lighting system—either a system using an electric track with wall-washing track fixtures or one using portable wall-washers.

Wall-grazing, on the other hand, is created by positioning fixtures 6 to 8 in. (15 to 20 cm) out from the wall. The light creates shadows under any surface change, bumps, or depressions. The surface changes should be intentional, not accidental. An owner can graze a wall with recessed downlights. A renter can use a portable track system behind a cornice board. Both owners and renters can graze a wall of draperies with valance lighting.

Since light calls attention to what it illuminates and can make things visible that seemed invisible before, be careful what is lighted.

When Not to Light One Wall

- When the surface of the wall is imperfect and has unintended flaws, like bumps or nicks. Light exaggerates these flaws. On the other hand, intended texture, like textured wallpaper or fabric, is not disadvantaged.
- When the interior space is so narrow that lighting along one side could make it appear lopsided or create other odd visual or architectural effects.

- When you do not want to call attention to the wall, for whatever reason.
- When the wall covering is a mirror or has a glossy finish.
- When the wall is divided into small segments with small windows or doors.
- When the ceiling is sloped along the special wall, making the wall height uneven.
- When the wall is opposite large, undraped windows or glass doors.

When to Light One Wall
- When the wall is finished differently from the other walls.

- When you want to spread a small amount of light a long way.
- When you want to make a small room seem larger.
- When you have a very deep-colored wall, which recedes from view.
- When the wall is uninterrupted.
- When draperies cover the major portion of the wall.
- When you want to shorten an interior space, such as the end of a long hallway.
- When you want to balance the other lighting in the room.

wall-washing

Wall-Washing for Owners

Owners can choose either recessed incandescent wall-washers or built-in fluorescent cornice lights. Both can put the amount of light needed on the wall. Both function on regular 120-volt household current.

RECESSED INCANDESCENT WALL-WASHERS

For rooms with 8-ft (2.4-m) ceilings where a soft, low level of illumination similar to the glow of many candles is desired, use recessed incandescent wall-washers in the smallest size with 30- to 50-watt reflector bulbs. These fixtures do not attract attention to themselves. They fit into ceiling joist spaces that are 6 in. (15 cm) deep. The bulbs direct the light forward, and the fixtures redirect it onto the wall. They illuminate about 5 ft (1.5 m) down the wall (the visible wall above furniture). The closer to the wall they are positioned, the higher up on the wall they will light. Consequently, for graphics or murals that have details all the way to the ceiling, install the fixtures between 1 ft to 1 ft 6 in. (.30 to .46 m) away from the wall. For wall coverings that have details further down, install the fixtures 2 ft to 2 ft 6 in. (0.61 to 0.76 m) from the wall.

Likewise, fixtures need to be spaced symmetrically. Install them the same distance from each other on center—from the center of one fixture to the center of another—in an equal distant pattern, dividing the leftover space at each end equally. Thus, on a 16-ft (4.9-m) wall, seven fixtures could be spaced 2 ft 3 in. (69 cm) apart with 1 ft 3 in. (38 cm) at each end. Be sure measurements are made from the center of one fixture to the center of another.

For building, provide wiring for the wall-washers when the electricity is roughed in. For relighting, recessing fixtures is possible if ceiling space permits, if wiring is accessible, and if fixtures can be secured to the ceiling material. In many of my relighting jobs, fixtures from one company have been adapted to accept the retaining clips from another in order to secure the fixture to the ceiling. Some companies are more

Wash a wall with little recessed wall-washers.

aware of relighting needs than others. Nonetheless, adaptations can be made.

The lighting effect is a series of gentle scallops are created at the top of the wall where the light from one fixture arches to meet the light from another. A dimmer is not necessary; the system allows three choices of bulbs for a little more or a little less light. When you want low-keyed illumination for quiet conversation, viewing television, or feet-up relaxing, use this system as the only light source. At other times when brighter light is desired, lamps and other sources should be available in the room.

Sample Electric Cost

On a 16-ft (4.9-m) wall with seven fixtures of 30 watts each, this wall-washing system requires a total of 210 watts—equal to one good desk lamp. If the lights are turned on every night for 4 hours and if the charge for electricity is 10¢ per kilowatt-hour, it costs $2.52 per month. The light will be effective and efficient.

For rooms with high ceilings, dark walls, or commercial spaces where a greater brilliance of light is desired, use recessed wall-washers with standard A, PAR, or reflector bulbs in higher wattages—75 to 300 watts. These wattages require fixtures two-thirds larger than the smallest size. They require 8 in. (20 cm) or more ceiling joist height. They pack a greater punch of light which can be scooped forward with a reflector or spread with a lens to create a smooth, scallop-free wash of light. The wall-washers that use standard A bulbs provide low-intensity illumination for spaces with 8-ft (2.4-m) ceilings. The PAR- and reflector-bulb types provide medium- to high-intensity illumination, depending upon the beam width, wattage used, and ceiling height.

Sample Electric Cost

On a 16-ft (4.9 m) wall with seven fixtures of 150 watts each, a wall-washing system requires a total of 1050 watts. If the lights are turned on every night for 4 hours and if the charge for electricity is 10¢ per kilowatt-hour, it costs $12.60 per month. The light will be spread along one wall and will be very much in view.

BUILT-IN FLUORESCENT CORNICE LIGHT

Fluorescent cornices are excellent for all styles of interiors and consume less energy. They are excellent for period-style interiors and traditional furnishings, which are normally difficult to combine with fluorescent lighting. They are

Wash a wall with a built-in fluorescent cornice light.

also sleek and tasteful in contemporary interiors.

A cornice light, if properly installed, confines all of the light and does not attract attention to itself. It focuses on the wall. This system can accommodate ceilings that slope away from a wall, provided the wall height is even. The fluorescent light goes about halfway down the wall, trailing off in brightness.

If other light sources in the room are incandescent, choose the kind of fluorescent light that puts out a compatible color of light. Otherwise, the colors will fight. Also, the color must flatter the occupants. Cool white, commonly used in offices, will not be flattering. Warm-white, warm-white deluxe, or prime-color fluorescent will. In addition, use all the same type color tubes in the cornice to provide a uniform color on the wall. If one tube color is changed, change them all.

Since the 4-ft (1.2-m) fluorescent fixture is the most common, use it. It is easy to obtain. But not every wall is divisible by four; another size might be needed too. For example, a 15-ft (4.6-m) wall could not accommodate only 4-ft (1.2-m) fixtures. Use three 4-ft (1.2-m) fixtures and one 3-ft (0.9-m) fixture, or some other combination of sizes that fit.

Be aware that only the 3-ft and 4-ft (0.9-m and 1.2-m) rapid-start fixtures can be dimmed. The instant or preheated fixtures cannot. However, dimming is usually not necessary, because fluorescent light is already low in wattage. Likewise, fluorescent fixtures do not usually create unpleasant brightness if the installation is designed well.

All fluorescent fixtures are distinguished by the quality of their ballast, but the quality is hard to determine. Rely on the rating of several organizations that test ballasts. The best ratings are a Class P for safety, a CBM certification for quality standards, and an A rating (on a scale of A to D) for the quietness.

A cornice system can illuminate evenly only a wall that is uniform in height. Therefore, the ceiling must be either flat or evenly sloped away from the wall. The farther the fixtures are installed from the wall, the farther down the wall the light will be thrown. In fact, for each 1 ft (0.3 m) out from the wall, the light will go down 4 ft (1.2 m). If you want light on the visible wall above furniture, install the fixtures 1 ft (0.3 m) out.

Place the fixtures end to end, within the limits of the fixture sizes. Normally, some empty space is left over, but keep it to a minimum and equalize it at each end. Or, if more than a minimum space is left, equalize it by separating each fixture by an inch or so, using up the extra inches. However, be aware that empty spaces between fixtures create shadows that destroy the smooth-wash effect.

Use this installation method:

1. Make a cornice board of hardboard or plywood faceboard, no less than 6 in. (15 cm) deep and as long as the wall being lighted.

2. If necessary to prevent the tubes from being seen from a usual position in the room, seated or standing, a narrow return or piece of wood can be added to the bottom of the board. In addition, the return could hold a diffuser or louver to further hide the tube. Neither is necessary in most installations.

3. Paint the inside of the board flat—not glossy—white.

4. Attach the board to the ceiling at least 6 in. (15 cm) out from the wall.

5. Mount the fixtures on the ceiling next to the board either on 1- by 3-in. (2.5- by 7.6-cm) wooden blocks or directly behind the board if decorative molding is to be attached on the outside at the top. Either way, the light cannot leak out at the top of the cornice board.

6. The outside of the cornice board can be painted, stained, covered with fabric or wallpaper, or plastered. Use whatever enhances the style of the room and blends with the wall covering.

7. Modify the measurements of the depth of the cornice board by the proportions of the board to the space, the view under the board, and other architectural considerations. Also, modify the measurements of distance from the wall by considering how far you want the light to be thrown down on the wall.

When building, a cornice system can be pre-planned and the wiring provided. The cornice board can be made to match the other woodwork in the space, especially if it is distinctive.

When relighting, a cornice system can be installed if wires are available either from access through the attic, from a hot switch, or from a baseboard receptacle. Skilled electricians often can reach wires in wooden frame construction, but not so often in other types of construction.

The light will spread as far as the fixtures are spread—ideally from end to end. More light will be at the top of the wall than at the bottom, but it will be bright and pleasing. The amount reflected into the room will be rich and glowing, sufficient to read the weather forecast, but not enough to read the whole newspaper.

Sample Electric Cost
A fluorescent system for a 16-ft (4.9-m) wall using four 4-ft (1.2-m) tubes of 40 watts each requires 160 watts for the tubes and 24 watts for the ballast. Used for 4 hours per night at 10¢ per kilowatt-hour, it costs $2.21 for a month. Fluorescent bulbs last up to 20 times as long as incandescent bulbs, and they spread light farther. In addition, they create a lot less heat. The overall saving—fewer watts for lighting and less cooling by fan or air conditioner in hot weather—would offset the slightly higher installation cost of the fluorescent system.

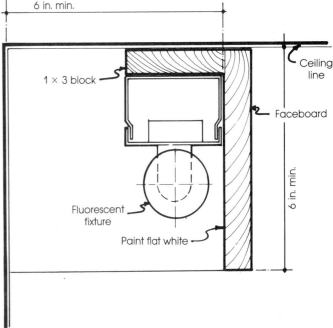

Cornice details.

6 in. min.

Ceiling line

1 × 3 block

Faceboard

Fluorescent fixture

Paint flat white

6 in. min.

Wall-Washing for Renters

Renters can install a temporary wall-washing system consisting of a track and track fixtures or portable wall-washers. Both become decorative elements in the space, especially in color and scale. They attract some attention to themselves, but they also put attention on the wall in the form of light.

TRACK AND TRACK FIXTURES

Track fixtures accept a range of bulb wattages for wall-washing from 50 to 300 watts. They utilize tubular (tungsten-halogen), PAR, standard A, and reflector bulbs. In residences, the 50- to 75-watt size can provide wall light graded from soft to shadow. The higher wattages—150 to 300 tungsten-halogen—in wall-wash flood fixtures provide smooth, bright light from top to bottom and corner to corner. They are especially good for residences with dark walls or high ceilings and for commercial spaces where brilliance, drama, and emphasis can be turned into dollars.

Track fixtures throw the light forward. They are adjustable and convenient for installation uncertainties—changes of mind, timid installation calculations, or refining the aiming angle because a ceiling slopes away from the wall. However, the purchaser pays for the ability to move the fixture.

Install track sections parallel to the wall and as long as the wall. Do not use a short track and aim the fixtures obliquely, trying to throw the light farther; it will be glaring and the light will be uneven. Use a track long enough to have each fixture equally spaced. The distance to the ends of the track from the last fixture should be equal to approximately half the distance between fixtures. Typically, for average-height ceilings (8 ft or 2.4 m), fixtures can be spaced between 2 and 3 ft (0.6 and 0.9 m) apart and the same distance from the wall. For higher ceilings, track distance and fixture spacing can be farther, but the farther back and the farther apart, the less the light will be received on the wall. Higher ceilings definitely require higher-wattage fixtures.

If chairs, sofas, or other seats are in front of the wall, make sure the light will not shine into the eyes of the person seated. If you are in doubt, position the track closer to the wall.

The wall will be washed smoothly with moderately intense light, which will be sufficient to illuminate dark wall coverings like wood paneling and deep tones of textured grass cloth. On paler walls, the amount of light re-

Wash a wall with track lights.

Renters can install a temporary wall-washing system of uplights.

flected should be cheerful and sufficient to enable one to enjoy activities with simple tasks, such as putting photos in the family album.

PORTABLE WALL-WASHERS

Portable wall-washers permit renters to take advantage of the benefits of wall-washing without disturbing the structure or the terms of the lease. They stand on the floor and connect with a cord and plug. However, they light from below. For these reasons, they are highly visible and make a definite impact. They are decorous for any interior that would not be disadvantaged by many floor fixtures—usually a contemporary or eclectic one.

Portable wall-washers are either canister-shaped, about 12 in. (30 cm) high and 6 in. (15 cm) wide, or a track fixture mounted on a weighted canopy base. Each must be equipped with a switch and must be turned on independently, unless the wall receptacles are wired in series (together) and can be controlled from one wall switch. A switch allows a dimmer to be installed. Use a reflector bulb in a 150-watt size, a 90-watt PAR, or a standard A bulb in 90- to 150-watt size.

All portables must be positioned evenly. Typically, the distance is about 3 ft (0.9 m) from the wall and about 3 or 4 ft (0.9 or 1.2 m) apart for the whole length of the wall, making the end

Portable uplight.

spaces about half the distance between the fixtures. Experiment with your wall to discover the best positions.

With these portables, the light originates from the floor. Therefore, the brightest part of the wall is at the bottom. Wall coverings containing details down to the floor can be well illuminated. Wall coverings with details farther up should have the fixtures positioned farther back. Floor fixtures can be hidden behind furniture in front of the wall.

Sample Electric Cost

On a 16-ft (4.9-m) wall, five portable wall-washers with five 100-watt bulbs, used 4 hours per night at a utility charge of 10¢ per kilowatt-hour, cost $6 per month.

wall-grazing

Wall-grazing is created by aiming light at a shallow angle to the wall. The fixtures must be positioned very close to the wall. The light enhances any texture, or other surface qualities, all of which should be intended. If the wall has unintended flaws, the light calls attention to them and they become predominant. Do not graze plaster or drywall that is only painted. Graze a wall covered with textured wallpaper, fabric, or masonry—particularly where the mortar has been removed from the edges (raked joints). A masonry wall stands up and sings with grazing light.

Certain requirements must be met to produce grazing light:

- The fixtures must be close to the wall.
- Incandescent reflector bulbs must be used. Standard A bulbs do not concentrate the light enough, and fluorescent is too shadow-free.

- The light bulbs must be well concealed, especially from view when seated.
- The fixtures must be close together.
- The fixtures must be out of sight and unimposing, either as recessed fixtures or track fixtures behind a cornice board.
- To produce the greatest amount of light with the least electricity, a recessed fixture must have a built-in reflector, and a cornice board must be painted flat white inside. Recessed fixtures without internal reflectors will graze a wall for a shorter distance, such as above the mantle over a fireplace.

When the fixture is positioned close to the wall, a pattern of scallops is created at the top of the wall where the scallop of one light meets the scallop of another. For some people, the scallops heighten the dramatic effect. Others disagree. If it is desired, the scallops can be virtually eliminated by positioning the fixtures very close to each other. However, the energy consumption and the heat are increased. In some installations, these disadvantages are outweighed by the look of the wall.

See for Yourself: Do You Object to Scallops?
1. Take two flashlights to a plain, pale-colored wall.
2. Light the flashlights and hold them about 12 in. (30 cm) apart, with your arm against the wall, one near your elbow and one in your hand.
3. Observe the scallops.
4. Note that the farther apart you hold the flashlights, the deeper the scallops become.

Wall-Grazing for Owners

Owners who are building and relighting can use recessed, multiplier-type downlights attached to ceiling junction boxes. The multiplier-type fixtures have internal reflectors that focus all the light and redirect it down. The reflector, sometimes called a cone reflector, is made of highly polished metal. Some reflectors are not well designed and show an image of the light bulb on the reflector. Choose one that does not. Inspect reflectors in your local showroom.

The greater the wattage, the more intense the shadows. However, good shadows come from 50-watt reflector bulbs if the fixtures are designed to control and direct the light and if the wall is not too dark. If it is dark, a 75-watt or higher watt bulb is necessary. Determine the color of the wall before buying the fixtures.

Install the fixtures 6 to 8 in. (15 to 20 cm) out from the wall and about 16 in. (41 cm) apart, evenly spaced. The further the spread, the deeper the scallops. The distance at the ends should be half the distance between the fixtures. Uneven spacing will be very apparent in wall-grazing. It will create uneven scallops and uneven light.

The system can be dimmed. Often, dimmers have been used where the initial wattage was too much for the space and the brightness had to be cut down to what it should have been in the first place. This practice is poor lighting design. Instead, dim to extend the life of the bulbs, especially if the installation is not in a convenient location to change bulbs. In addition, dimming changes the color temperature of light and emphasizes the red end of the spectrum. Therefore, dimmed light could be used to enhance, for example, red tones on a brick wall.

When you are building, plan for sufficient ceiling depth for wall-grazing fixtures. When you are relighting, the depth must be there already, and wiring must be gained through an attic or by pulling wires from elsewhere. If ceiling wiring is not obtainable, owners will have to use the temporary wall-grazing methods suggested for renters.

Wall grazing makes the wall texture very visible, enriching and enlivening the special wall. The light will be sufficient for conversing, listening to music, or viewing television.

Sample Electric Cost
Light for an 8-ft (2.4-m) wall using six fixtures of 50 watts each for 4 hours per night at 10¢ per kilowatt-hour costs $3.60 per month. The amount is no more than would be required for two table lamps, and yet two table lamps could not light a whole wall.

Wall-Grazing for Renters

Renters do not have the options that owners have to recess fixtures. Nonetheless, they can have dramatic grazing light by two means—either track fixtures hooked up with a cord and plug, or bare bulbs on a track behind a cornice board. Track fixtures have their own shielding, and can be installed to point almost straight down along the wall. Bare bulbs on a track need a cornice board to shield them from view.

To make a cornice board, follow the directions for the built-in fluorescent cornice light in this chapter. However, mount the board at least 8 in. (20 cm) and up to 12 in. (30 cm) away from the wall, depending upon the wattage of the bulb (higher wattage bulbs are bigger). Bulbs of 75 watts or less require cornice boards to be 8 in. (20 cm) from the wall; 150-watt bulbs require cornice boards to be 12 in. (30 cm) from the wall. When ceilings are 9 ft (2.7 m) high or less, suitable installation is 8 in. (20 cm) out, and suitable wattage is 75 or less.

Make the board 12½ in. (32 cm) deep with a 3-in. (8-cm) return or less at the bottom. The track should be mounted on the inside of the board. The outside of the board can be trimmed, painted, or otherwise finished to fit and coordinate with your wall. Install a cornice board as long as the wall to be grazed, and adjust the track fixture to point down.

Use the simplest track fixture—a socket for the bulb—and equip it with a spot bulb. These fixtures are functional-looking and must be hidden. Each bulb should be 10 to 12 in. (25 to 30 cm) from the other and 6 in. (15 cm) in from the end. However, different spacing or type of bulb

Graze a textured wall with recessed lights.

might be more satisfactory for your wall. See for yourself by checking the effects.

Choose the wattage of the bulb in relationship to the color of the wall and height of the ceiling. In average-height rooms with pale- to medium-color walls, use 30- to 50-watt bulbs. For more impact, use 50- to 75-watt bulbs, particularly the new 65-watt PAR that gives the same amount of useful light as the 150-watt one. Dark walls and ceilings higher than 8 ft (2.4 m) need more wattage to illuminate them effectively. This is not to say that dark walls should never be used. On the contrary, dark walls are rich and attractive, but they require more light.

The precise alignment and wattages need not be determined before installation. Changes of spacing and wattage can be made afterward. Often changes are required in both commercial and residential spaces.

The wall will be brighter at the top and softly lighted down the wall, depending upon the wattage chosen. For example, the effect of using the 75-watt spot is a strong stroke of light touching the wall all the way down. It puts out enough light to let you glance at a magazine or find your shoes easily, but not enough to read a fifth carbon copy of an order.

Sample Electric Cost

A cornice system using eight 50-watt fixtures along an 8-ft (2.4-m) wall for 4 hours at 10¢ per kilowatt-hour costs $4.80 per month. On the other hand, a custom-made system equipped with 75-watt spots, spaced 12 in. (30 cm) apart costs $7.20 per month. Even with the higher wattages, the yearly additional cost of electricity for a wall-grazing system is about the same as four tanks of gas for the family car or one for the office delivery truck.

grazing window walls and draperies with a valance

Often, large windows take up a whole wall or part of a wall. Sometimes these windows are covered with draperies, which are usually closed at night. The draperies become a major element in the space, contributing color and texture. Lighting the draperies not only reflects light from the wall but also bounces it off the ceiling, giving the space greater illumination. Light the drapes with a valance. Valances can be used easily with sloped ceilings because they are installed well below the ceiling line and are not affected by an uneven wall.

A valance is basically like a cornice but is installed 10 in. (25 cm) or more down from the ceiling. The faceboard can be made the same depth as a cornice board, but it is attached to the wall. (If the board were installed at the ceiling, it would be called a cornice board.)

Certain requirements must be followed to install valance lighting:

- The faceboard must be installed a minimum of 10 in. (25 cm) from the ceiling to let enough light out.
- The light bulb or tube must be at least 2 in. (5 cm) away from the faceboard and the draperies to dissipate the heat.
- The depth of the faceboard must be at least 6 in. (15 cm), but usually needs to be deeper to conceal the light source from seated people.

Valance lighting on draperies bounces light into the space.

- The board must be painted flat white inside to reflect as much light as possible.
- The board must be at least 4½ in. (11 cm) out from the wall for clearance.
- The light bulb or tube must be at least 2 in. (5 cm) up from the bottom of the faceboard to prevent seeing it.
- The fluorescent fixture should be installed on 2-by 3-in. (5- by 8-cm) wood blocking to eliminate escaping light.
- The light should be smoothly spread for the whole length of the drapes.
- The light source can be fluorescent (preferable) or low-wattage incandescent.

The light from a valance will be soft and appealing. It will spread as wide as the fixtures are spread and will appear above and below the board. It will permit short-term, easy-to-see activities, such as setting a table or eating an informal meal. Or, in a commercial space, it will give a broad area of light so that there are no dark corners.

Valance light can be either fluorescent or incandescent. A fluorescent valance system consists of fluorescent light fixtures behind a valance board. (Install it the same way as in a built-in fluorescent cornice light described earlier in this chapter.) An incandescent valance system is composed of incandescent fixtures installed behind a valance board. The valance board can be ready-made or built especially for the space. (See the directions on wall-grazing for renters earlier in this chapter.)

Owners who are building or relighting should use fluorescent valance lighting. On the other hand, renters can use incandescent track fixtures temporarily hooked up with a cord and plug and hidden behind a ready-made or custom-made faceboard. The ready-made board limits the wattage to 50, good for pale to medium-color draperies. The custom-made board

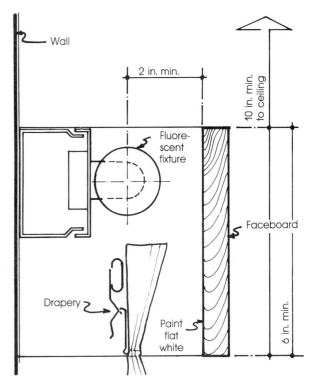

Valance details.

can accommodate higher wattages, if necessary, but 75 watts would be enough for most dark draperies in spaces lighted by other sources.

If the draperies do not cover the whole wall end to end, the face board must have end pieces to obscure the light fixtures and complete the valance. The face board can be held to the wall with ½-in. (1.3-cm) metal angle brackets or wooden brackets. The board can be finished to harmonize or contrast with the interior design of the space.

The amount of illumination ranges from bright to soft, depending upon the color and texture of the draperies, the wattage of the bulbs, and the light in the rest of the space.

downlights
for settings

6

Downlighting is one of the best ways to light. Downlights can create many settings for homes or businesses that are not obtainable by any other type of lighting. Downlights can create sparkle, and can carve out, expand, and punctuate a space. They can provide a lot of or a little light that harmonizes or contrasts. With downlights the ceiling is not bright; surfaces—tabletops, walls, furniture, and floors—are bright. Consequently, surfaces attract your attention, not an uninteresting ceiling.

See for Yourself: Do Surfaces Attract Attention?
1. In a dark room, turn on one lamp with a translucent shade.
2. Observe that your attention is drawn to the lampshade.
3. Open the bottom and top of a box big enough to cover the whole lamp, yet allow the light to go to the ceiling.
4. Observe that your attention is drawn to the ceiling surface and you are unaware of the lamp at all.

Some downlights are well-engineered fixtures. Choose them to get the most light for your money. Some downlights put the light straight down; some put it to the side. How the fixture looks does not always indicate the kind of light it produces. Do not try to guess what it does.

In theaters, settings are created by skillful placement of well-engineered lighting equipment. In commercial spaces, settings are created by downlights to sell products. Likewise, settings can be created in residences to emphasize, soothe, or brighten. What kind of setting do you want?

sprinkle a horizontal surface with sparkle

Create a brilliantly sparkling tabletop with light. It is attention getting and cheerful. The colors and objects on the table become vivid and appealing. Contrast the setting to a more subdued background. Make the downlight bright.

The setting draws people to it. The closer the light is confined to the tabletop, the more intimate it becomes. This effect is good at home as well as at the store.

Two ceiling downlights or a downlight in a

chandelier can provide this kind of setting. **PAR** or reflector bulbs will do it best. In most residential cases, a new energy-saving 65-watt or a 75-watt PAR flood from the ceiling or a 50-watt reflector from a chandelier will be sufficient. For more <u>brilliance</u>, add more w<u>attage</u>. For narrow defined be<u>ams</u> of light, use <u>spots</u>.

Sample Electric Cost
The electrical cost at 10¢ per kilowatt-hour for two ceiling fixtures with 65-watt bulbs for 4 hours per night for a month would be $1.56.

On the other hand, for commercial spaces, the amount of illumination required depends on the amount of light around it. The amount of illumination must be three times as bright as the surroundings. Consequently, 500 footcandles might be required for featured tabletop displays. Check the display area with a light meter and determine how much light is needed.

carve out space
within a space

A space can be defined by downlights. The light need not be brilliant, but the surrounding light must be dimmer in order not to loose the integrity of the carved-out space. Brighter illumination could be further away if it did not interfere.

For example, downlight, focused on comfortable furniture inviting you to sink-in, visually organizes the seating area with consistent light. The light creates seemingly three-dimensional but penetrable walls. It is soothing and yet sociable. The illuminated area seems to be alone and suggests you sit back, put your feet up, and forget the cares of the day.

In residences for a softly illuminated area, use small reflector downlights of 30 to 50 watts. For a vividly illuminated area, use 75- to 150-watt PAR's, or the equivalent in the newer energy-saving bulbs (65 or 90 watts). Position the downlights over the front or the back edge of the seats, <u>never over the center.</u> Downlight would produce ghoulish shadows on faces of the people seated.

A softly illuminated seating area of one

Carve out a space within a space with downlights.

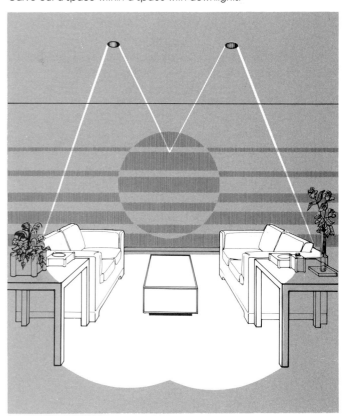

small and two large sofas, and a coffee table would require three 30-watt downlights. Position the fixtures over the front edge of the cushions at the ends of the large sofas and over the middle of the small sofa. In addition, position a 50-watt downlight over the coffee table to fill the interior of the area.

Sample Electric Cost
The electrical cost for one 50-watt and three 30-watt downlights for 4 hours per night for a month at 10¢ per kilowatt-hour would be $1.68.

---

expand a space
with widespread light

A room can be expanded by light, exploding the space and fostering an atmosphere that encourages people to move around freely. Light reveals the richness of the whole space. It is stimulating. Several fixtures are required to spread the light uniformly. The amount of light from each fixture should be the same, and the fixtures should be spaced evenly. When choosing a fixture, consider personal preference about the fixture's aesthetics, and coordinate its light-delivery capabilities. Wide-beam downlights can do the job. Fluorescent fixtures can also work, but they create distractingly bright patterns of light on the ceiling. Unfortunately, they are overused in commercial spaces. Parabolic wedge louvers on the fixtures will make them less obvious. However, substitute downlights in some spaces and provide visual relief from the rows of fluorescent fixtures that march across most commercial ceilings.

When using incandescent downlights to expand a space, create uniform light by positioning the fixtures evenly throughout the space. The distance between fixtures can be determined easily with spacing ratios. However, if a specific footcandle level must be delivered, more complicated calculations will have to be made. (See calculations in Chapter 21.) To easily determine spacing, either follow the manufacturers' spacing ratio, or position the fixtures apart by the same distance they are from the floor. For example, an 8-ft (2.4-m) ceiling should have fixtures 8 ft (2.4 m) apart. If fixtures are too far apart, the light becomes scattered, chopping up the space rather than unifying it. If they are spaced correctly, the light is uniform.

Manufacturers indicate spacing ratios in their technical catalogs. They are the maximum distances their fixtures can be apart and still deliver uniform light. A spacing ratio multiplied by the height above the surface to be lighted (usually the floor, possibly a tabletop) equals the space between fixtures. The space between a fixture and the wall should be half the distance between the fixtures.

Rule of Thumb for Spacing Fixtures
Spacing = Spacing ratio × height above the surface to be lighted

---

Spacing ratios range from 0.3 to 1.4. The greater the ratio, the farther apart the fixtures can be installed. For example, in a room with 8-ft (2.4-m) ceilings, fixtures with a spacing ratio of 1 should be centered 8 ft (2.4 m) apart; fixtures with a ratio of 1.4 should be centered 11 ft 2 in.

Expand a space with a widespread light.

(3.4 m) apart. Of course, the actual installation distance must be modified by the size of the room, because room sizes cannot always be divided evenly by spacing ratios. As you modify the spacing, be aware that the closer a fixture is to a wall, the more likely a scallop of light will be formed on that wall. Sometimes the scallop can illuminate something you do not want seen, disturbing an interior setting. Determine any disadvantages before installation.

punctuate a distinctive small area with a pool of satiny light

A small bay window in a residence or a low-rise commercial structure can be illuminated. From the outside, the lighted bay window greets and says goodbye. At night such a light can be a nightlight, giving a reassuring glimmer. When the room is lighted and full of people, the downlight balances the other illumination. The downlight need not be bright. If the surrounding light is not overpowering, 30 watts can be bright enough. If the surrounding light is bright, 50 or 75 watts may be necessary.

Sample Electric Cost

Only 36¢ worth of electricity is used for a 30-watt downlight operating 4 hours per night for a whole month at 10¢ per kilowatt-hour.

Many other settings can be formed with downlights. Determine what is to be emphasized and how vivid it should be. Of course, the more vivid it is, the more attention it will draw. But not all settings require brightness, nor should they be bright. Remember, vivid light is stimulating; soft light is soothing. Decide which mood you want.

In addition, the mood can be changed with a dimmer—a device to reduce the light output of the bulb. As the incandescent light dims it also turns more golden in color and gives the appearance of being warmer, enhancing the red colors in the space, but sometimes making other colors appear muddy.

From the outside, a lighted bay window greets and says goodbye.

Punctuate a distinctive small area with a pool of satiny light.

effects

The lighting effects for settings range from broad to focused, from harmonizing to contrasting, and from soft to vivid. When the whole room is illuminated, a social, gregarious atmosphere is created. When an intimate space is illuminated and the rest of the space is dark, people are drawn to the lighted space. When the walls are illuminated, a self-contained, soothing atmosphere is created. The effect of downlights depends not only on the engineering of the fixture, but also on the type of incandescent bulb used. Some downlights are engineered with reflectors to concentrate the light; others spread the light. Fixtures that put light straight down produce wide, medium, or narrow lighted areas. The appearance of the fixture does not always indicate the width of the light it produces. For example, a pinhole looks like it produces a narrow beam of light, but it does not. It produces a medium beam if the bulb is correct and properly adjusted. A narrow beam is produced by an open fixture equipped with a very narrow spot bulb. Do not try to guess the amount of light a downlight produces; consult the manufacturer's catalog, a knowledgeable salesperson, or a lighting consultant.

fixtures

The engineered features of the fixture are internal or external. Internal features are reflectors and adjustments. Reflectors redirect, spread, or concentrate the light from within the fixture. Some reflectors are better designed than others. The best not only cut off the view of the bulb but also do not reflect the bulb's image. Internal adjustments are found only in well-engineered downlights. Their function is to aim the light to accent or to compensate for the slope of a ceiling.

External features are lenses, diffusers, baffles, and louvers. Lenses cover the opening of the fixture and bend the light in some way. A Fresnel lens concentrates the light. A spread lens redirects it to the left and right. A prismatic lens directs the light down, but it can also be bright. To make the best use of a lens, the bulb needs to be the exact wattage specified by the manufacturer. Then the light will be precisely focused. The engineering of a downlight can be ruined by the wrong bulb.

Diffusers obscure the light bulb and scatter the light. Opal diffusers are the best to use. When they are the only sources of light, they are bright, attract attention, and in some situations are distracting. Smooth out the brightness with additional light sources in the space.

Baffles (black grooves in the inner surface of the fixture's opening) and louvers (metalwork over the aperture) also reduce aperture brightness. Some louvers are unidirectional, intended to throw the light in the direction they are pointed. However, these are not as effective as internal reflectors.

All downlights can be described in terms of how they distribute the light: wide, medium, or narrow beam; adjustable accent; wall-washer; framing projector; or downlight/wall washer.

A wide-beam downlight is generally used to illuminate a large area. A medium-beam downlight is used to illuminate an area where the activity and the objects to be viewed are important—a tabletop, a seating area, or around other furniture. (On the other hand, the space becomes more important and the activity becomes less important when walls are illuminated.) A narrow-beam downlight is used to

Reprinted by permission of Lightolier.

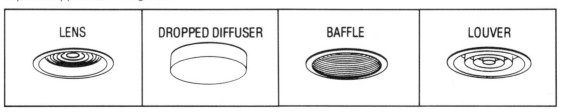

| LENS | DROPPED DIFFUSER | BAFFLE | LOUVER |

Wide.

Medium.

Narrow.

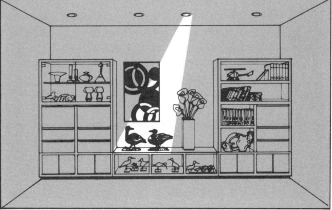

Adjustable accenting.

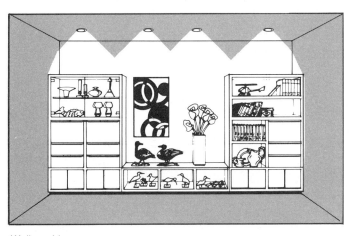

Wall-washing.

Framing projector.

Downlight/wall-washer.

highlight special areas or objects. In rooms with high ceilings, many narrow-beam downlights can illuminate a whole room. Also, both wall-washers and adjustable accent fixtures are engineered to put light to one side. (Wall-washers are discussed in Chapter 5.) Adjustable accent fixtures can point to the side, or they can shoot light straight down from a sloping ceiling, compensating for the slope. Some are internally and some are externally adjustable. Likewise, a framing projector shoots light either straight down or to the side. It produces a sharp-edged beam of light. (Consult Chapter 8.) Finally, a downlight/wall-washer is very useful for lighting two different surfaces at one time. It permits some light to fall down, as well as some to be directed to one or two walls, or even to a corner.

Some downlights are not well engineered. They are essentially tin cans with sockets. They can trap light. When a whole room is to be lighted, more tin can–type fixtures are required than are well-engineered fixtures. In addition, well-engineered fixtures are not necessarily the most expensive in the long run. Fewer of them are needed to light a space and fewer bulbs are needed, which means replacing fewer bulbs and using less total wattage. Each manufacturer makes several quality lines, from well-engineered to standard, which are expensive to budget-priced. Because of the competition in lighting, some budget-priced fixtures are reasonably engineered and are suitable for many residences and commercial spaces with 8-ft (2.4-m) ceilings. High or sloped ceilings definitely require more expensive fixtures and a higher quality light delivery.

bulbs

Downlights accommodate a wide range of incandescent bulbs, from 25 to 500 watts. Many downlights accept more than one wattage or type of bulb. These fixtures are internally adjustable to ensure that the bulbs are in the most favorable positions to deliver the light. If they are not properly adjusted, light is wasted within the fixtures themselves.

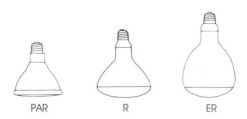

Reprinted by permission of Lightolier.

In general, narrow beams of light come from spot bulbs, either PAR's or reflectors, and from fixtures designed to deliver narrowly from standard A bulbs. Wide beams come from standard A bulbs as well as PAR or reflector (R) flood bulbs. Elliptical reflector (ER) bulbs deliver narrower beams than floods, but wider beams than spots. The 75-watt ER fits in all 150-watt R- or PAR-size fixtures. In many situations, the 75 ER gives as much light as is needed. It delivers about the same amount of light as a 150-watt R, but in a slightly smaller area. In addition, the ER is also available in a 150-watt size. The new energy-saving bulbs, such as the 65-, 90-, or 120-watt PAR or the 67- or 90-watt standard A, reduce energy consumption but also deliver less lumens (the unit of light coming from the bulb), anywhere from 60 to 320 lumens less. This reduction, however, is considered barely noticeable, and the amount of light remains more constant over the life of the bulb. Test the bulbs in your setting.

When choosing between standard A, PAR, or R bulbs, there is a trade-off. Standard A's produce more light per watt, but all sides of the bulb emit light which may get trapped in the fixture. Standard A's are cheaper to buy, but they burn out quicker. PAR's, R's, and ER's are more expensive, but they last longer. Consider the tungsten-halogen PAR, which lasts even longer and is suitable for commercial spaces or for any fixture that is hard to reach, even in residential spaces.

Each fixture is manufactured to use the specified bulb efficiently. Reflector and PAR bulbs produce heat, and so the fixture must have a ceramic socket. Sockets for standard A bulbs do not have to be ceramic. Follow the manufacturers' specifications for bulb type, and do not try to outwit the engineering of a fixture.

installation

All downlights must be connected to the electricity, either through a ceiling junction box or by way of a track. Through a junction box, fixtures are installed recessed, surface-mounted, or semirecessed. The fixtures are hooked up to the electricity within the ceiling. Recessed fixtures require sufficient space to put the fixtures above the finished ceiling. The recessed fixtures are the least conspicuous, and they enhance the spontaneous effect of the setting. In wooden frame construction, the ceiling joists usually are 16 in. (41 cm) on center, leaving available 14 in. (36 cm) in width and 10 in. (25 cm) in depth. However, if the available space is not sufficient, semirecessed (dropped partially below the ceiling) or surface-mounted downlights might be possible.

The National Underwriters' Code requires that recessed downlights either be thermally rated—engineered to permit covering with ceiling insulation—or have a thermal switch in case they get too hot. Before thermal protection requirements, all downlights were installed 3 ft (0.9 m) away from insulation. During that time, one of my clients asked me to determine how much heat and air would be lost with ceiling downlights if the insulation had to be 3 ft (.9 m) away. After consultation with another university and the State Energy Office, I gave him a determination. We did not have a chance to verify the calculations because he bought a house and did not build. Now, thermal-rated downlights can be covered with insulation, and they will prevent loss of heat or air conditioning.

In concrete and steel construction, tracks can be surface-mounted and plugged into a baseboard receptacle, with track fixtures installed as downlights. In all construction types, tracks also can be hooked up to ceiling junction boxes. In commercial spaces with suspended ceilings, downlights are easy to install.

Owners can install recessed fixtures in new construction if wires are provided. When relighting, owners must have an electric junction box in the ceiling, or the wires can be brought in through the attic. In addition, the electricity must be controlled by a wall switch. A dimmer can be installed to give greater flexibility of lighting control in both residential and commercial spaces.

On the other hand, renters must use a track with a cord and plug for downlights, as do owners of concrete and steel structures. Track parts are highly visible. Therefore, the cord could be obscured in the corners of the wall and ceiling on its way to the baseboard receptacle, to make the installation as inconspicuous as possible. If required, a dimmer can be installed on the cord.

After most track fixtures are installed, they can be adjusted up, down, and side to side. Some are less adjustable than others. Determine the fixture limitations before you buy, and you will be able to achieve many kinds of settings.

to change your mind about lighting

7

Life is full of changes: People change their minds, demands change around them, and their bodies change. Modifications need to be made to accommodate these changes. Lighting can be modified to accommodate some of these changes.

people change their minds

Often people like to change the style, color, and arrangement of their surroundings. Customarily, change requires a great effort, not to mention expense. However, changing lighting can be as simple as a click if the lighting fixture is attached to a track canopy. At the same time, it can be reasonably priced.

Examples

- Renters, formerly stuck with looking at a chandelier chosen by someone else, can convert for a small investment the original chandelier and a chandelier of their choice to track adapters. The original chandelier can be put away in the closet, while their chandelier graces the dining room. When the lease is up, they can put back the original and be off with no further electrician's bill and no landlord hassle.
- People who change their minds and want two chandeliers for the same room can do so. They can adapt two chandeliers to track adapters. Clip one into a track canopy one day and replace it with another chandelier the next day. For instance, hang an antique brass turn-of-the-century fixture on Tuesday and a silvery chrome chandelier on Sunday. The chandelier can be changed to satisfy mood or mind.

Instructions

A canopy, from the accessory part of track catalogs, is installed over a ceiling junction box. The canopy accepts one fixture at a time, either a track fixture made by the same manufacturer or a chandelier or pendant made by any manufacturer and fitted with the appropriate adapter. Track manufacturers have limitations for their adapters. The limitations are the components of the fixture, the diameter of the cord, the stem's thread size, or the total weight. Before purchasing, check the catalogs for the limitations of the track, and coordinate with the components of the chandelier. Track canopies and adapters are inexpensively priced. A canopy and two adapters usually cost about the same as a moderately priced restaurant dinner. Unlike a dinner, the effects of lighting are enjoyable night after night.

Put up one chandelier one day . . .

. . . and change it the next.

demands change around people

Sometimes demands require that a space to be used in a different way, either temporarily or permanently. When demands change, track components can accommodate these changes.

Examples

- When the bloom leaves the Western store business, and when hats, boots, and fringed jackets give way to an aerobic dance studio, the bright merchandising spots can be replaced with G bulbs around the mirrored walls, and they can even be wired to blink with the music.
- When dinner is to be served buffet style and the dining room table must be moved out of the way, a chandelier, formerly hung permanently over the table but now converted to a track with an adapter, can be replaced. A downlight can be put up for the duration of the buffet, and no one will bump his or her head.
- When the bridge club is coming and the dining room is to be filled with card tables, an adapted chandelier can be replaced with an adapted fluorescent ceiling fixture. The fixture will throw light broadly over the tables. After the card games, the dining table and chandelier can go back.

- When a relative comes to stay and a living space gets converted to a bed space, the electric track, formerly with accent lights only, can be converted to hold accent lights and pendants for the dresser or for bedside tables.

Limitations

The possibilities of change are numerous. Many residential chandeliers and pendants can be track-adapted and moved from side to side. Some track manufacturers make adapters for their own ceiling surface-mounted fixtures. In addition, old track fixtures can be replaced with new ones in a different style, as long as the track and the fixtures are made by the same manufacturer.

There are a few limitations. Track fixtures only fit tracks made by the same manufacturer. Commercial-size chandeliers are too heavy for track adaptation. When lengthening a track, only add another track from the same manufacturer. A fixture in a track canopy cannot be moved from side to side.

When a relative comes to stay,
a former family room . . .

. . . can become a bedroom.

people's bodies change

Less Light

Some people are uncomfortable in bright light. The reasons vary from vanity and self-consciousness to illness and other physical reasons that can cause the eyes to be sensitive to light. Further, if you feel that some lights are too bright, it may be neither vanity nor illness. The lighting might be poorly designed. If so, you must first determine the cause; then you can decide on the cure.

CAUSE: CONTRAST

Is there too much contrast between the light and the darkness of the rest of the space?

CURE: OTHER SOURCES

If so, add additional light sources and reduce the contrast.

See for Yourself: How Much Contrast Is There?
1. Bring several lamps into a room.
2. See how the original light appears less bright.

CAUSE: BRIGHT BULB

Is the brightness caused not by contrast, but by seeing the light bulb?

CURE 1: ADJUST

If so, and if the fixture is adjustable one, adjusted it to block the view of the bulb. Often such fixtures are installed with little regard for glare. Track lights are the most likely to offend because installers are unaware of the critical rule of thumb for track fixtures.

Rule of Thumb for Track Fixtures
Never position a track fixture to allow a person to look into it and see the bulb.

CURE 2: BLOCK BULB

If the track fixture is positioned with the bulb in view and if the fixture is not adjustable, obscure the light bulb by blocking the view to the bulb with a louver, a shade, or a baffle. Louvers are sometimes available from the manufacturers of the fixtures. Louvers obscure the bulb from many angles. Get or make a suitable louver. For example, if a table lamp blasts light up the stairs, blinding those coming down, get a perforated metal circle and attach it to the top of the shade. At other times, when obscuring devices are not available for the fixtures, they can be custom-made. The basic supplies are in do-it-yourself building supply, electrical, and hardware stores, or a sheetmetal shop can make them. However, never enclose an open downlight with a lens unless the manufacturer specifies in the catalog that it is enclosable.

Shades either block the light source completely (opaque) or diffuse it (translucent). For example, if the track bulb is bare, get "barn doors" from track-lighting accessories and hang them on the bulb. For a lamp, purchase an opaque shade that conceals the bulb.

A baffle can be made of anything that blocks the view of the fixture—fire-retardant cloth, wood, or composition boards. Arrange them to reflect light but at the same time obscure the source.

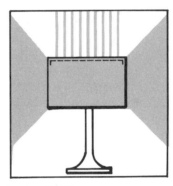

Lamp with diffuser.

CURE 3: SHORTER BULB

If the light bulb is visible, possibly it is too long for the fixture. If so, get a shorter bulb. Some identical wattages are obtainable in long or short lengths. Check with your local electric supply store. On the other hand, some fixtures are designed to have the face of the R bulb stick out. Do not install a shorter bulb in these fixtures because heat might build up inside and be a hazard.

CAUSE: TOO MUCH LIGHT

Is the brightness caused by the amount of light?

CURE: REDUCE THE LIGHT

If so, reduce the wattage of the bulb or reduce its light-producing capabilities. If it is an incandescent fixture, reduce the wattage by purchasing a lower-wattage bulb, or install a dimmer on a wall switch or on the cord. Dimmers save electricity and extend the life of the incandescent bulb by reducing the voltage. Reducing the voltage 50 percent extends the bulb life up to 16 times; reducing it 20 percent extends the bulb life up to four times. Longer bulb life and greater flexibility of settings are gained with dimmers. Both incandescent and fluorescent dimmers are available. Otherwise, a fluorescent tube can be wrapped with white, heatproof electrical tape (glass tape) found at electric supply stores for a nonvariable, permanent dimmer. The tape holds back the light given off by the tube. The more it is covered with tape, the more light it holds back. Usually, wrapping the tape with a 1-in. (2.5-cm) space between the strips is satisfactory. This method permanently dims the light. It is not as variable as a real dimmer that controls the intensity smoothly for different amounts at different times.

More Light

Perhaps the interior surfaces of a room have been changed to a dark color, absorbing more light. Or, as in the case of merchandising, a display may have been changed and the details smaller, requiring more light. Maybe more light is needed because the space is used by older people. As people grow older, they require more light, although—at whatever age—some people just need more light than others. Consequently, other light sources need to be added or higher wattages need to be used.

See for Yourself: What Is the Value of Light?
1. Take the phone book to a dim hallway and try to read the numbers.
2. Take the phone book to a lamp with a 150- or 200-watt bulb.
3. Observe how much better you can see the numbers.

CAUSE 1: DARK COLORS

Are interior surfaces—floors, wall, furniture—deep, dark colors?

CAUSE 2: OLDER EYES

Are you older than you were when you installed the lighting, or has someone older come to live with you?

CURE: FOR INCANDESCENT LIGHT

If you have incandescent fixtures, increase the amount of light by one of the following methods:

- Increase the wattage up to but not higher than the maximum specified by the manufacturer, which is usually indicated on the inside of the fixture.
- If the fixture is capable of handling it, change at the same wattage from a reflector bulb to a PAR bulb; the output will be about two times brighter. (Fixture capability is obtainable from the manufacturer's catalog or from a knowledgeable manufacturer's representative, an electrician, or a lighting consultant.)
- If possible, change from a flood to a spot bulb, either a reflector or a PAR-type. A spot light puts out 150 percent more light than a flood light, but in a 60 percent smaller area. Unfortunately, more light is traded for a smaller area.
- If the ceiling fixture is recessed and covered with a lens, get an open trim from the manufacturer of the fixture. Most recessed fixtures have alternative trims. Lighting consultants and lighting showroom employees can help find them.

Reflector bulb.   PAR bulb.

All increases in the amount of light are relative because the eye does not perceive increases accurately. In most cases, doubling the amount of light is seen as a 50 percent increase. Likewise, increasing the amount of light and not increasing the size of the source of light concentrates the brightness. Therefore, if more light is needed but brightness is not, the amount and the size of the source must be considered. For example, the wattage of a bulb in a large Japanese lantern could be increased easily without increasing the concentrated brightness, because the large surface would disperse the light. A small lantern would not. On the other hand, a ceiling fixture previously equipped with a 60-watt bulb, but changed to a 150-watt one, would appear brighter. Whether the light would be too bright depends upon your preferences. If you want less brightness but more light, choose the largest surface to disperse the light. Excessive brightness and more light need not go hand in hand.

If you have an electric track and want more light, increase the number of track fixtures without exceeding the maximum capacity for the wattage of the track (around 1,800 watts) or for the electric circuit to which it is wired. You or an electrician can determine the capacity of the circuit.

See for Yourself: What Is the Capacity of the Electric Circuit?

1. Add up all the watts to be consumed in one circuit and divide by 120.
2. The resulting number is total amperage; it must not exceed the current capacity marked on the circuit breaker or fuse.
3. If it exceeds current capacity, the fuse will blow or the circuit breaker will trip, resulting in no electricity on that circuit.

All increases in lighting draw some electricity. However, one or two extra lights are unlikely to overload most circuits, because lighting consumes so little electricity. But if lighting is on the same circuit as small appliances—blow dryers or color televisions—it is another matter. If you are in doubt, check with an electrician.

CURE: FOR FLUORESCENT LIGHT

If you have fluorescent fixtures and available space:

- Install another fluorescent fixture.
- Or install a double-tube fixture, making a total of three tubes. (The second tube increases the amount of light by 85 percent; the third, by 70 percent—for a total of 155 percent more light from the fixture.)

Change your lighting either temporarily or permanently, as you require.

light
for cooks and
noncooks

8

People who like to cook need more places lighted in a kitchen than do people who do not cook. However, noncooks need minimal lighting at the work counter, at the sink, and at the cooking and eating surfaces. Most kitchens have only a center ceiling light, which gives unsatisfactory light. With a center ceiling fixture, kitchen counters are in shadow wherever you stand. Effective kitchen lighting illuminates the work being performed. Light at work locations reinforces the intended use of the location and focuses attention there. Light the kitchen by first determining the locations for lighting and by identifying where a fixture can be installed—on upper cabinets, above upper cabinets, from the ceiling, or on a shelf. Fixtures for noncooks are quickly installed; fixtures for cooks require more elaborate installation, but provide comfortable and well-integrated lighting. Some cooks may want to choose fixtures listed for noncooks, for their own reasons. After determining lighting locations, determine the amount of light needed from each fixture. The amount required varies according to the color of the kitchen surfaces; the other light sources available, including daylight; and the condition of the user's eyes. The best method of determining how much light is needed is to test it yourself.

See for Yourself: How Much Light Is Needed?

If you are considering fluorescent light:

1. Determine the length of the tube that fits.
2. Take an incandescent bulb that is at least three but not more than four times the wattage of the fluorescent tube. (For example, to test light from a 40-watt fluorescent tube, hold up a 150-watt incandescent bulb.)
3. Hold the incandescent bulb where you would install the fluorescent.
4. Observe whether the light delivered is sufficient; do not look at the bulb itself or the size of the area lighted, just the amount of light. If it is not sufficient, choose a fixture that has two fluorescent tubes in it.

If you are considering incandescent light:

1. Identify the amount of wattage the manufacturer specifies for the fixture under consideration.
2. Take an incandescent bulb of that wattage and hold it where you would install the fixture.
3. Observe whether the amount of light delivered is sufficient. If not, add more light by choosing a fixture that can deliver more, or choose additional fixtures for other locations.

At the Countertop

To illuminate work locations at a countertop, fixtures can be hung under the upper cabinet or at the ceiling, or they can be stretched between the upper cabinets. Noncooks will probably have only one counter to light; cooks will probably have several.

CABINET ABOVE THE COUNTERTOP

If there is an upper cabinet above the countertop, attach a fluorescent fixture to it. Get one that is about two-thirds the length of the counter. For instance, if the counter is 4 ft (1.2 m) long, the fixture can be 3 ft (.09 m) long. A ready-made undercabinet fluorescent fixture installs quickly and conveniently with a cord and plug, which is ideal for renters and noncooks. The switch is on the fixture or the cord. Owners can also use a ready-made fixture but should install it permanently, directly into a junction box behind the wall, not temporarily by plugging into a wall receptacle.

With ready-made fixtures, decide ahead whether the light source should be obscured. Some fixtures have a plastic diffuser to spread the light but not hide the tube. Others block the tube from both the seated or standing viewer. These fixtures could be a miniature shielded track, a shielded fixture, or an inexpensive fluorescent strip that can be thrown away. The inexpensive strip can be installed to create its own shield. In addition, upper cabinets can provide structural shielding themselves; that is, the door or lower part of the cabinet can block the tube from view. However, custom-made fluorescent lighting can be more distinctive.

Custom-made fluorescent lighting is built into the upper cabinet. It consists of a tube holder (strip) and replaceable fluorescent tube. Do not use the throw-away type fixture. For the best lighting distribution, the strip should be installed in the front part of the upper cabinet. But if the backsplash or countertop is pale-colored and not glossy, the strip can be installed at the back. When relighting, if the upper cabinets have shielding or if shielding can be built in, both owners and renters can custom-install fluorescent strips.

When building, if the cabinets are to be custom-made, owners can design the upper cabinets to shield the fluorescent strips with:

Put light where you work.

- a door that hangs below the bottom shelf
- a structural support behind which a strip can be placed
- a recessed lower shelf that creates a place for the strip. (Inside the cabinet, the recessed shelf creates storage for short objects—canned goods, glasses, and cups.)

Sometimes a diffuser or louver is needed with a fluorescent strip. The cabinet needs to be 3½ in. (8.9 cm) deep to hold a diffuser or a louver. A diffuser spreads the light and obscures the source. It can be prismatic or opal. A louver controls light in case the tube might be seen by someone seated in the same room or in an adjacent room. A louver can be an egg-crate grid, a parabolic wedge, or some other type.

It is not always necessary to have custom-made cabinets to accommodate fluorescent strips. Some ready-made cabinets can incorporate them with minor modifications. However, modify them before hanging on the wall; afterward it is more difficult.

A door can hide the fixture.

A recessed shelf holds a fluorescent light.

NO CABINET ABOVE THE COUNTERTOP

When there is no cabinet above the countertop, illumination can come from a recessed, surface-mounted, or suspended ceiling fixture. A recessed fixture is the least obtrusive. Regardless of which type you want, choose a fixture with a lens or diffuser. A lens either spreads or concentrates the light; a diffuser scatters the light and obscures the bulb. A dropped diffuser spreads more light than a flat one. Do not use an open ceiling fixture—a downlight with no diffuser or lens—unless extra brilliance is required. An open fixture gets dirty quickly, and if it is used alone, the light can be harsh. Ceiling fixtures are far away from the surfaces that need light, and they therefore require more wattage to get the light down. Use at least 40 fluorescent watts or around 75 incandescent watts or more, according to individual needs.

Also, if there is no cabinet above the counter, illumination can come from a pendant or a suspended fixture. The fixture can be decorative, and it can provide a repeatable interior design feature if coordinated with the same fixture elsewhere. Hang a pendant high enough to allow for headroom. Choose one that transmits light through the sides as well as down. Otherwise, the light might be too harsh if it is the only source.

Finally, if there are two side cabinets, illumination can come from a lighted shelf stretched between them. The shelf produces light and holds kitchen items. The light can go up and down, spreading over more than just the counter top. Three benefits are received from one shelf. The shelf can be made of wood, glass, or plastic. Use a fluorescent strip as a light source, and get one as wide as the shelf. Mount it under the front edge. If possible, use a rapid-start fluorescent 36-in. (91-cm) or 48-in. (122-cm) strip. It turns on instantly and is, therefore, the most suitable.

At the Sink

The most frequently used place in the kitchen is the sink. Light it well. The amount of light should be enough to help distinguish between a dirty dish and a clean one. The fixture can be attached to the upper cabinet or to the ceiling. If there is a cornice board between two upper cabinets at the ceiling, a fixture can be mounted behind it.

MOUNTED ON THE CEILING
BEHIND A CORNICE BOARD

If a fluorescent light source is to be used, mount it behind a cornice board. The strip should be as long as possible. For instance, if the width of the board is 50 in. (127 cm), a 48-in. (122-cm) fluorescent strip would fit. The type of illumination produced is shadow-free, but not very intense.

Lighted shelves also hold kitchen items.

The amount from one fluorescent tube might not be enough; two or three might be required. (Directions for installation are in Chapter 5.)

If an incandescent light source is to be used, a 75-watt incandescent reflector bulb, either regular (R-30) or elliptical (ER-30), is excellent. The elliptical bulb puts out a medium-intense area of light, equivalent to the light of a 150-watt flood bulb, but it consumes less electricity. The area it lights is the size of most sinks. Install the bulb either in a very inexpensive porcelain socket, if the cornice board is deep enough to shield the 6⅜-in. (16-cm) bulb, or in a downlight that is surface-mounted or recessed. Normally, downlights designed for 150-watt bulbs are sufficiently deep.

MOUNTED WITHOUT A CORNICE BOARD

If a fluorescent source is to be used, a single- or double-tube fixture can be recessed or surface-mounted. If an incandescent source is to be used, a recessed or surface-mounted downlight can hold a reflector bulb, equivalent to 75 watts or more, giving a bright pool of light. If a decorative accent needs to be continued, use a ceiling-hung pendant. Before purchasing a pendant be sure that it provides enough light. Test for yourself the maximum wattage specified by the manufacturer.

At the Cooking Surface

Even noncooks need to see when the water boils. Cooks need to see a lot more. Consequently, cooking surfaces (ranges, stovetops, or built-in units) need light. If a hood over the cooking surface has a light, use it with the wattage specified by the manufacturer. Do not reduce the wattage unless you are particularly sensitive to light. If you do not have a hood, install a shielded fixture.

If upper cabinets are not available, a lighted shelf, a pendant, or a ceiling fixture can light the cooking surface. The amount is more critical for cooks who need to distinguish between browned and burned food. But both cooks and noncooks will want a minimum of 65 to 75 incandescent watts, or 30 fluorescent watts. Use fluorescent light whenever possible; it is cooler, and the cooking surface produces enough of its own heat. Special infrared incandescent bulbs can keep foods warm elsewhere.

At the Eating Surface

An eating surface may be a counter, a bar, a peninsula, or a table. Noncooks are more likely to eat in the kitchen than are cooks. Eating requires illumination, unless you like eating in the dark. Eating surfaces thrive on light. The more direct and intense it is, the more appealing the food looks. Incandescent bulbs produce the most direct and intense light.

Direct and intense light can come from any incandescent fixture that puts light down hard on the eating surface—open downlights (either surface-mounted or recessed), downlight pendants, and downlights in chandeliers. For a long rectangular eating surface, several pendants spaced about 20 in. (51 cm) apart or a linear chandelier (a library or billiard fixture) accent the architecture of the surface. However, in a small eating area a less conspicuous fixture is more aesthetically pleasing, such as ceiling-mounted light, either recessed or on the surface. Remember, the light should call attention to the eating surface, not to the fixture.

FRAMING PROJECTOR

The most distinctive and intense light for eating surfaces is produced by a framing projector. In theaters it is used to light stage objects precisely. In eating areas, it can illuminate just the eating surface and nothing else. It is capable of cutting the edges of a light beam. When fully open, it can light an area about 60 in. by 60 in. (152 cm by 152 cm). When closed down, it can light a smaller area. The light can be confined to shine brilliantly on all table settings and food, enhancing colors and appetites. The light source is a tungsten-halogen incandescent bulb. It requires careful handling, but burns brighter longer than other bulbs. The fixture needs readjusting each time the bulb is replaced. These requirements are fine for some people and terrible for others. The framing projector can be either recessed in the ceiling or track-mounted. Thus, both owners and renters both have equal opportunity for using this precise optical instrument.

However, states with residential Lighting Power Budget Limits (limits on the amount of watts allowable in new construction) eliminate incandescent light as a choice. Fluorescent light sources must be used in order to stay within the prescribed limits. Although they are neither direct nor intense, fluorescent lights can be pleasing. Choose warm-white deluxe or prime-color tubes, which render colors better. Prime color costs about three times more but is worth considering. Eating surfaces are used 365 days each year, and the additional cost would be less than half a cent a day for the life of the tube.

Fluorescent lights can be suspended over the eating surface, either in ready-made fixtures or built into a canopy. The ready-made fixtures are slick-looking and accent any nontraditional interior. A canopy is a wooden frame, open above and covered with a diffuser below. The light from the fluorescent strips bounces off the ceiling and down through the diffuser. Make the canopy proportional to your eating surface and within the sizes of fluorescent strips (2 ft to 8 ft, or 0.6 m to 2.4 m). Suspend it 10 in. (25 cm) from the ceiling in an average-height room, lower in a higher ceilinged room. Make the depth of the canopy at least 6 in. (15 cm). The distance between the strips should be one and a half times the depth of the frame. The distance from a strip to the side of the frame should be half the distance between the strips. Therefore, if the frame is 6 in. (15 cm) deep, the strips should be 9 in. (23 cm) apart.

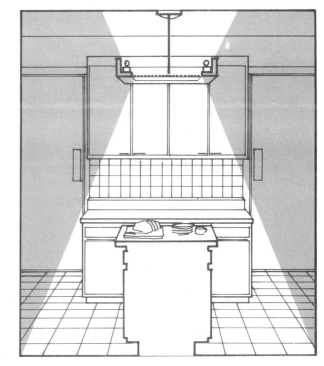

A canopy bounces light up and down.

light for the room itself

Sometimes the room can be illuminated sufficiently by the light from the sink, the stove, and the other work locations, but sometimes it cannot. Cooks will want to see easily into upper and lower cabinets, and they may require more light sources. Several alternatives are available.

- If the upper cabinets do not go all the way to the ceiling, cove lighting can be installed on cabinet tops with excellent results. Cove lighting consists of fluorescent strips that bounce light off the ceiling and spread it around the room. Collectibles or usable kitchen objects can be displayed above the cabinets and lighted either from in front or from behind. Choose whichever lighting looks best and installs easiest. (See the directions in Chapter 9.)
- Fluorescent lights can be built into a dropped soffit above and in front of the upper cabinets, or they can be built as wall brackets in the same position. (See the directions in Chapter 9.)
- A center ceiling fixture can be either surface-mounted or recessed. A recessed fixture with a dropped lens spreads the light on the ceiling; a flat lens sends the light down. Some surface-mounted fixtures permit the light to come through the sides of the fixture; some do not.
- If the fixture can be seen from another room, choose one that does not send light through the sides. When used alone, too much glare would be created, because the light would contrast with the darkness of the room.

- A center ceiling fixture illuminates the center of the room, which in most kitchens is empty. Therefore, the light should not be overly bright and attract attention. Work locations should be brighter. Countertops always look better than floors. If the center is not empty and contains an island or a peninsula, install the ceiling fixture directly above it.
- In small windowless kitchens luminous ceilings

Cabinets support cove lights.

are excellent. They provide broadly spread room lighting from low-energy-consuming fluorescent strips. They must be augmented by undercabinet lights; otherwise, the cook creates his or her own shadow. (See the directions in Chapter 9.)

Luminous ceilings for small kitchens.

other lighting

Other lighting is available for kitchens.

- A warming light, an infrared reflector bulb, can keep food warm before serving. It can be built into an upper cabinet. These bulbs are particularly useful at passthroughs. A 125-watt infrared bulb is sufficient. It heats up ard cools down fast. The heat it produces is clean and concentrated.
- Lighted cabinets with glass or plastic cabinet doors transmit the light. (See directions for furniture lighting in Chapter 11.)
- Luminous panels can be used between the countertop and the upper cabinet as the backsplash or between two upper cabinets as a cornice. Both provide not only light but also a decorative ambience. The panels can be glass or plastic. For the backsplash position, choose panel colors that do not impart a distinct hue to the light. Red, green, or blue, for example, would make food look odd. For the cornice position, choose any hue. The light below will still be white, and the light transmitted through the panel will be colorful without causing interference. In the backsplash position, a panel should be illuminated by fluorescent strips. Build a box for them. Place them at the top and the bottom, or at the left and right sides, 4 in. (10 cm) back from the edges so that they will not show. To obtain even light, the box should be one-sixth as deep as the distance between the strips. For example, if the strips are 21 in. (53 cm) apart, the box depth could be 3½ in. (8.9 cm). Paint the inside of the box flat white. Hinge the panel so that the tubes can be replaced when they burn out, usually not sooner than three

A stained-glass luminous panel between upper cabinets.

years. Moreover, plan how you want to control the luminous panel; a wall switch is the most convenient to use.

using multiple fixtures

All of the fixtures in a kitchen do not need to be on at the same time. However, I had one client who insisted that all the kitchen lights be on one switch, all on or all off. She did not want to switch lights at different locations, no matter how convenient it was. Contrary to this feeling, the advantage of multiple lighting is that lights can be used alone or in combination, where and when needed. Multiple lighting is energy-efficient, since unused light can be turned off. Light up your life in the kitchen—easily and satisfactorily.

light
to dress and
bathe by

To clothe and bathe ourselves we need mirrors, tubs or showers, and closets. Sometimes we need ironing boards and exercise areas. All of these require light. Light for mirrors must be on the person, not the mirror. Light for tubs, showers, and closets must be in the tubs, showers, or closets. Light for exercise areas should not be on the person, but should be reflected from somewhere else in the room.

at mirrors

In the Bedroom

Mirrors above a dresser or dressing table are commonly lighted by lamps. The design of the lamp and shade is not just a matter of aesthetics but of function. If the shade is opaque, narrow at the top, and broad at the bottom, it squeezes the light down, illuminating the top of the dresser, not your face. Choose a shade that transmits light and is 2 in. (5 cm) taller than the bulb. If the lamp is tall enough, it will illuminate your face well; if it is too short, glare gets into your eyes. It almost cannot be too tall, unless the bottom of the shade is above your eye level.

Rule of Thumb for Lamps at a Mirror

Place a pair of lamps about 3 ft (0.9 meters) apart on either side of the mirror.

Experiment with various size bulbs to get the exact amount of light desired. Instead of an incandescent bulb, consider using a fluorescent adapter, which holds a circular fluorescent tube and adapts to the screw-base socket, or use the newer bulb-shaped fluorescent sources, which have excellent light output and a screw-base. Fluorescent sources emit more than three times as much light as incandescent and use at the least one-third less electricity.

Other bedroom mirror light sources could be wall-hung, ceiling-recessed or surface-mounted fixtures, two pendants, or a chandelier. (Refer to Chapter 3.)

A Full-Length Mirror

Ideally, the fixture that lights a full-length mirror should be out of sight but should illuminate the person fully. This is very hard to do. On the ceiling, position a fixture close to the wall, and include a diffuser to soften shadows under the eyebrows, nose, and chin. Usually full-length mirrors are located on doors or short walls where ceiling fixtures fit, both architecturally and aesthetically. Architecturally, they are out of the way; aesthetically, they do not attract unnecessary attention to themselves. On the wall, position one fixture above the mirror or two on the sides.

In the Bathroom

The usual location for a bathroom mirror is on the wall above the sink. The size of the mirror determines which lighting fixtures should be used, and the status of ownership determines how the fixtures can be installed. Owners can more easily install fixtures connected to the electricity in the wall or ceiling. Renters, on the other hand, must use plug-in wall fixtures. Either way, the fixtures should control the brightness and distribute the light.

SMALL MIRRORS

If the mirror is small, you can use light fixtures in several places.

- *On the ceiling and on the wall at each side of the mirror.* On the ceiling, the fixture could be surface-mounted or recessed. Either one will illuminate the head and shoulders. A recessed fixture does not conflict with any interior style. A surface-mounted ceiling fixture, on the other hand, needs to be chosen with care to blend well. On the wall, the fixtures should be 28 to 36 in. (71 to 91 cm) apart on the sides and centered about 60 in. (152 cm) from the floor, if the person will be standing. Remember, the ultimate aim is to have soft, localized light at the mirror.
- *On the wall at each side and above the mirror.* Use either incandescent or fluorescent wall fixtures. Incandescent fixtures are highly stylized. Some emulate historical styles and some are like theater dressing-room lights, surrounding the mirror on two or four sides. Use the lowest wattage suitable for visual requirements in incandescent fixtures. Fluorescent sources are preferable beside the mirror because they provide good distribution for little energy. For example, the fluorescent sources on each side of the mirror could be 20 watts and that in the ceiling 40 watts, for a total of 95 watts, including 15 percent extra for the ballast. Incandescent sources, on the other hand, would require about three times as much electricity to do the same job. Choose incandescent fixtures with opal glass diffusers, because they obscure the bulb.
- *Theater lights (small incandescent bulbs in rows).* Theater lights are popular for small mirrors. They are usually placed along the sides of a mirror. The bulbs should be 15 watts or less, rather than 40 watts or more, even though the manufacturer indicates that the larger ones are us-

Light on both sides and above the mirror reflects on your face and neck.

Theater lights beside a mirror.

For a dramatic effect,
fluorescent light behind a mirror.

disperse the light by reflection. Large mirrors look appropriate with large built-in lighting. Fluorescent fixtures are large sources of light, can be surface-mounted, and are architecturally suited to being built in. Owners may want to build a dropped soffit or a lighted wall bracket to hold fluorescent tubes. A soffit directs light down, reflecting it from the countertop and the mirror to the face. Likewise, the bracket directs it down, and it also directs it up if desired to flood the ceiling with light. A ready-made surface-mounted ceiling fixture can be used by owners who choose not to build in lighting or by renters who cannot build in. A ceiling fixture over the mirror provides light in the same way as a soffit or bracket. Renters can connect them to a track hooked up to the electricity through a ceiling junction box where a ceiling fixture is now located.

- *Dropped soffit.* A dropped soffit is a boxed-in area of a ceiling dropped 8 to 12 in. (20 to 30 cm) down. Dropped soffits look best above a vanity top and should be the same length (left to right), but not the same depth (front to back) as the vanity. The soffit can be 12 to 18 in. (30 to 46 cm) deep and should be able to hold a diffuser to scatter the light. Use two singles or a double fluorescent strip, unless the room has medium- to dark-colored walls or unless your eyes require more light. If so, use three or four fluorescent strips. Paint the inside of the soffit flat white to reflect as much light as possible. The vertical surface of the soffit should be finished to blend with the bathroom walls, and the bottom should blend with the ceiling. For example, a soffit could be built of sheetrock, plastered, painted off-white on the bottom to match the ceiling, and wallpapered on the outside to match the walls.

able. For the least glare, choose the coated bulb. If clear bulbs are preferred, use other light sources in the room and dilute the glare of the bulbs. The wattage of each bulb and the number of bulbs in the fixture determine the brightness. In general, many low-wattage bulbs mean less glare; few high-wattage bulbs mean more glare. Theater lights can be joined end to end and can also be connected at right angles. Consequently, the total wattage used might be more than the electrical circuit can carry or the fixture's internal wiring can handle. Check on these details before joining many together.

- *Behind the mirror.* For a dramatic and decorative effect in guest bathrooms, both residential and commercial, where mirrors are used casually rather than seriously, fluorescent light sources behind the mirror are unexcelled. Mount the mirror on a wooden frame 4 in. (10 cm) smaller than the dimensions of the mirror and 4 in. (10 cm) deep. Attach fluorescent strips to the frame, at the top, the bottom, and the two sides. The mirror will be surrounded by a halo of light and appear to float. This effect is particularly flattering for dressing rooms that are intended to thrill the customer with his or her image.

LARGE MIRRORS

If a mirror is 3 ft (0.9 m) wide or more, it helps

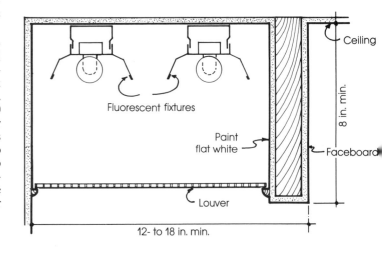

Soffit details.

A dropped soffit puts light down.

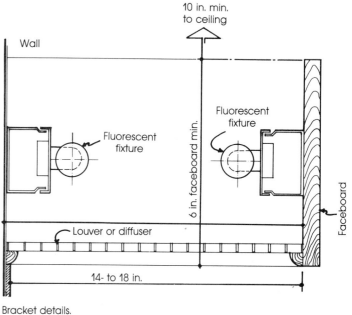

Bracket details.

A bracket light enhances
with light up and down.

- *Bracket.* A bracket is a cantilevered shelf holding lights. Like a soffit, it looks best above a
vanity top. It should be the same length as the
top, be at least 10 in. (25 cm) down from the
ceiling, to allow adequate reflection, and be 14
to 18 in. (36 to 46 cm) deep. The face board
should be at least 6 in. (15 cm) high and covered
with any material that enhances the interior de-
cor—wood, tiles, wallpaper, fabric, or paint.
Use two single fluorescent tubes to light the
bracket, one on the inside of the faceboard at
the top and one on the wall. The distance from
the end of the strip to the end of the bracket,
ideally, should be 6 in. (15 cm), in order to have
smooth lighting. Paint the inside of the faceboard flat white. Use a diffuser at the bottom.
Use white perforated metal on the top, covering
just the fluorescent tube at the wall or covering
both tubes for less light on the ceiling.

Individual differences, such as individual preferences, eye conditions, and interior colors, play
a big part in choosing the type and amount of
lighting for mirrors. Some people prefer theater
lights beside the mirror. Some have preferences
about the color of the light. Incandescent light
enhances colors in all faces. Fluorescent can do
so also, if warm-white deluxe or prime-color
tubes are used. Such tubes produce more red,

reflecting a truer skin color than do cool-white tubes. However, people who work in offices may want cool-white to apply makeup, since their office light is the same color. (Some portable makeup mirrors are made to light both ways.) Also, older users require more light—between two-thirds and twice more than others. Consequently, more wattage is required.

Different interior colors also affect the color of the light. To this point, one South Florida resident wrote to my newspaper column inquiring why her skin tone appeared different in two bathrooms equipped with the same kind of theater lights at the mirrors. In one room her skin tone looked true; in the other her skin appeared sallow. The reason was the color of the walls—a sunshine yellow—reflecting an unflattering hue on her. Never cover a large area in any space with a color that does not flatter you. The light, whatever the source, will pick up a tint of that color and, as a result, reflect that color on you.

protective wall
receptacles near sinks

Wall receptacles near sinks should be the protective type—ground-fault circuit interrupters. They guard against hazards of shocks originating in electrical appliances such as hair dryers and toothbrushes. (Shocks from the electric wiring in the wall will be controlled at the circuit breaker in the main power panel.) A ground-fault interrupter will shut off the electricity quickly when the flow is interrupted, even just a bit, preventing a shock. Electricity can be interrupted by faulty or worn-out appliance wiring or by getting the wires wet. These protective wall receptacles are available at local electric supply stores in a single or double size. They fit into the same space as the old receptacle. In many areas, building codes require them for new construction when a receptacle is located

Protective wall receptacle.

within 8 ft (2.4 m) of a kitchen or bathroom sink. People would do well to install them in older homes, particularly in bathrooms where teenagers use electrical appliances.

in closets

Light is mandatory in closets for inspecting, selecting, and finding clothes. Consequently, all closets 10 or more sq ft (0.9 sq. m) should have a light. Choose a closet light from one of these:

- a recessed downlight with a 25-watt reflector bulb
- a porcelain socket mounted on the inside above the door header with a standard A bulb 12 in. (30 cm) away from anything combustible
- a surface-mounted fluorescent strip on the ceiling two-thirds as long as the closet. The strip can be enhanced with a reflector behind the bulb. Such a system beautifully backlights the slats in a louvered door. Hallways lined with

louvered doors can be lighted entirely this way at night. Control the strips with a wall switch.

Closet lights can be controlled by one of three methods, in order of their expensiveness from cheapest to the most expensive—a pull chain directly on the fixture, a wall switch, or an automatic switch activated by the door opening or closing.

Sometimes closets have built-in dressers and mirrors. If so, illuminate them with theater lights equipped with 15-watt bulbs or less, so that the closet area will not be glaring or be hot from unnecessarily high wattage bulbs.

at the ironing board

Many of my lighting clients claim that they do not own an iron. Nevertheless, for those who do, ironing needs strong directional light to reveal wrinkles—incandescent, not fluorescent. Just being able to see the fabric reasonably well is not sufficient. Often clothes are ironed under low indirect light, and the wrinkles remain. They appear later in brighter directional light— on the way to the car, or outside while waiting for a taxi. Then it is too late.

Light from an oblique angle can reveal the wrinkles while the iron is in hand. To light the board minimally, use a 75-watt reflector bulb, either in the ceiling or in an adjustable floor lamp, and a fixture elsewhere in the room. Use higher wattages if you iron more than minimally, such as ironing what you sew.

over tubs and showers

Tubs or showers require a fixture, and most usually do not have one. The fixture must be watertight and should be recessed in the ceiling. The light aids in washing, grooming, and reading instructions on labels. Many of my clients who initially are somewhat unconvinced, cheer resoundingly after installing a tub or shower light, and wonder how they got along without one before.

Sometimes a tub is a special bathroom feature. An owner with a bathroom that has access to the roof can have a skylight to illuminate the tub during the daytime and have light fixtures in the skylight to illuminate it at night. (See Chapter 17.) On the other hand, a luminous ceiling can substitute for a skylight where a skylight is not possible. A luminous ceiling provides a large surface of light. Use fluorescent strips as the light source, and space the strips evenly behind a sheet of translucent plastic (a diffuser). Hold the plastic in place by a ceiling system consisting of a box the size of the ceiling over the tub (minimally 6 in. or 15 cm deep) and a T-bar grid the size of the diffusing material. The deeper the box, the more uniform the brightness of the light on the diffusing material on the ceiling. But whatever the depth, position the fluorescent strips apart by a distance equal to one and a half times the depth of the box. Allow half that distance to the edge of the box. Therefore, if the box is 6 in. (15 cm) deep, the fluorescent fixtures should be 9 in. (23 cm) apart and 4½ in. (11 cm) from the edge of the box on each side.

If the tub or shower area is embellished with plants, a luminous ceiling can keep the plants healthy, provided it is turned on for 14 hours a day and equipped with cool-white tubes. The area will be very appealing both day and night, and the temptation will be to take more leisurely baths or showers.

Sample Electric Cost
A 40-watt fluorescent light, including ballast, can be on for 14 hours per day for $1.93 per month, if electricity costs 10¢ per kilowatt-hour.

in exercise areas

Some people exercise at home and some exercise in commercial spaces. Exercise spaces need general room illumination, not lights at specific locations. One form of general illumination is a ceiling fixture that distributes the light, and is itself not overbright. These qualities are controlled by fixture design and diffuser. If the fixture design permits the light to spill through the sides as well as down, it spreads the light. If the diffuser is or poor quality, it will be too bright and the light source will be visible through it. Opal or ceramic-enameled diffusers are the best for incandescent fixtures. But if the fixture is small and overwatted, even a good diffuser can not overcome the annoying glare.

For fluorescent fixtures, diffusers that are

A tub with a luminous ceiling keeps plants alive.

rated to have a high visual comfort probability (70 or more) are the best. Fixture manufacturers indicate this rating as VCP. The better the VCP rating, the more comfortable the lighted diffuser appears. Comfort is always a lighting objective.

Many exercises are done lying on one's back, facing the ceiling. Therefore, do not position the fixtures directly over an exercise bench or mat location. If mirrors are included in the exercise space, be aware that the fixture will reflect in the mirror. The most glare-free lighting for an exercise space with mirrors is cove lighting. It is indirect. It is composed of a cove board on two or more walls, with fluorescent tubes behind it. Install the cove board at least 10 in. (25 cm) down from the ceiling to bounce as much light off the ceiling as possible. The ceiling then becomes a large light source softly reflecting in the mirrors, not glaring. This technique is

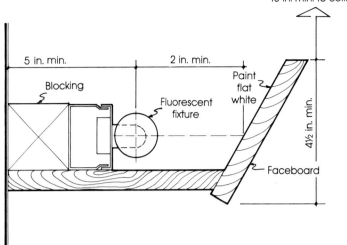

Cove details.

particularly suitable for commercial exercise spaces, where the patron's degree of comfort is critical for business.

accent
your paintings
with
low voltage
10

At night, an oil painting fades from view.

Turning on lamps does not help.

A painting is a special purchase. It is expensive and is proudly displayed on the wall, at home or at the office. It is visible, but yet not visible. At night, like the rest of the walls, the painting fades away forgotten. Turning on lamps does not help. The painting needs its own light. The artist used light to create it; but hung without light, the painting loses color, detail, texture, and beauty. The light can be dramatic if it illuminates the painting only; the painting appears to be self-illuminated. Less dramatically, the light can illuminate the painting and some of the wall. Or the whole wall, including the painting, can be lighted. The choice depends upon what effect is desired.

Accent a painting with light and consume very little electricity by using low-voltage lighting. Low-voltage operates at 5.5 or 12 volts rather than at regular line voltage—120 volts. Low-voltage lighting has several features. It provides an accurately controlled beam of light, a broad selection of beam widths, and a greater intensity of light for each watt consumed. Also, it is produced by a small fixture. The benefits, therefore, are less electricity consumed, less wasted light, and less heat. In hot weather or in crowded rooms, it does not add much to the heat load. Likewise, less heat is produced that could fade the pigments in paintings, especially in commercial and public spaces where paintings are illuminated for a long time. Museums use low-voltage bulbs to minimize fading.

A low-voltage fixture can be recessed, surface-mounted, or clipped into a track. The fixture must be adjustable in order to aim the light at the painting. Recessed fixtures normally are adjusted internally. Surface- or track-mounted fixtures are adjusted externally.

Low-voltage recessed fixtures can fit where standard sizes cannot. They are inconspicuous. Some have built-in transformers. Surface-mounted fixtures do not, and they can be used with simple wiring. Thus, they are easy to use when relighting an existing structure. Surface-mounted fixtures accept only regular reflector bulbs. Surface-mounted and track fixtures adjust the most vertically. However, some are more adjustable than others.

A track fixture can hold a heat filter to further reduce radiant heat. If you do not mind seeing the hardware on the ceiling, track equipment is a good choice for lighting paintings. If you do mind seeing the hardware, use a recessed low-voltage fixture.

track equipment

Track equipment looks theatrical. It consists of an electrified holder, a low-voltage fixture, a low-voltage bulb, and a connection mechanism for the electricity. A connector mechanism is either an electric junction-box cover or a cord for plugging into a baseboard receptacle. An electric junction-box cover can be at the end or anywhere along the track. The junction-box connection is permanent; the cord-set connection is temporary and less attractive.

Electrified Holder

The electrified holder is either a track for several fixtures or a canopy for one. Electric tracks are available in 2-, 4-, 8-, and 12-ft (0.6-, 1.2-, 2.4-, and 3.6-m) lengths and some can be cut shorter on the job. They are available in single to quadruple circuits. In residences, single circuits are used primarily; in commercial spaces, double, triple, and quadruple circuits are used when many items are accentuated at one time, or when changes of accenting are needed by the flick of a switch. Choose the number of circuits needed depending on how the track is to be used.

- If you want to light just the painting and are likely to move it on that wall, choose a single-circuit track and one fixture.
- On the other hand, if you want to light just the painting and never move it, choose a canopy and a track fixture. A canopy accepts a single track fixture and permits the fixture to adjust up, down and around, but not slide side to side. It creates a surface-mounted low-voltage fixture out of track components. The canopy is round, about 5 in. (13 cm) in diameter. It is installed directly into a ceiling junction box and wired to a wall switch. A canopy accepts any low-voltage fixture.
- If you want to light the painting and the wall at the same time, choose a single-circuit track and several fixtures.
- If you want to light the painting sometimes and the wall at other times, choose a double-circuit track, several track fixtures, and two switches.

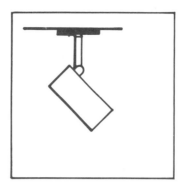

Canopy track light.

A canopy cover can substitute for an electric track.

Both circuits "on" in a two-circuit system.

Low-Voltage Fixtures

All low-voltage fixtures need transformers to reduce the regular line voltage. The transformer is either built into the fixture or installed separately from the fixture—called a remote transformer. Any transformer requires approximately 20 percent additional electricity to function. Low-voltage track fixtures can contain their own transformers.

Low-voltage track fixtures are made in black, white, brass, aluminum, or bronze. Choose a color to match the ceiling and reduce the visual impact, or to contrast and create a visual image. Either way, purchase all fixtures from the same manufacturer. Components are not interchangeable.

Low-Voltage Bulbs

A low-voltage fixture requires a low-voltage bulb. Three types are available. The first type is a well-engineered parabolic reflector (PAR) similar to a car headlight. The second type is a well-engineered reflector—a multifaceted reflector (MR-16) is similar to a projector bulb. The third type is a standard small reflector (R-14). These bulbs last 66 percent longer than table-lamp bulbs (standard A).

PAR   MR   R

Reprinted by permission of Lightolier.

Low-voltage bulbs produce a beam of light that is precise, allowing an exact determination of the area to be lighted. A PAR bulb has beam widths of very narrow spot (up to 20°) to very wide flood (80°), in wattages of 25 or 50. In some circumstances, the area of distinguishable light from the very wide flood appears smaller than the wide flood. At each wattage, the wider the beam the less intense the light is. The PAR bulb can operate at 5.5 volts and puts out a pin-point of light for objects far away.

The small reflector bulb has a spot or a flood beam in 15 to 25 watts. The multifaceted reflector bulb has beam widths of narrow spot, narrow, and regular flood. It concentrates the light for outstanding brilliance and lasts longer. It is made in 20-, 50-, and 75-watt sizes. The creative effects of all these bulbs are almost limitless.

Connection Mechanism

OWNERS

Owners can permanently install a track lighting system. A recessed track makes the installation look slick. A recessed track can be plastered right in the ceiling or cut into a suspended tile ceiling. Either way, the track is out of sight. Owners planning to build may want to specify the method of installation. On the other hand, owners planning to relight might not be able to recess a track. They can surface mount a track and connect it to the electricity, if access to electric wires is possible within the ceiling. Access is required for both surface-mounted and recessed track installations, and so is a wall switch. Access can be gained by one of four methods:

- removing a ceiling fixture and using the wires in the junction box
- pulling the wires from a junction box to another place on the ceiling
- pulling wires from a baseboard receptacle up through the ceiling
- rewiring above the ceiling in the attic from the main circuit breaker

An electrician can determine which method is possible. If access is not available or permanent installation is not desirable, owners must install tracks with cords and plugs, like renters.

RENTERS AND OTHERS

People who rent their space, short-term owners, and mobile-home owners must light their special paintings with a surface-mounted track system that connects to the electricity with a cord and plug. The advantage is that it is portable. Most tracks and some canopies can be ordered with a cord, usually about 15 ft (4.6 m) long, and a plug making them as convenient as a table lamp. Obscure the cord by taking it straight from the track to the nearest wall, proceed along in the ceiling-wall corner to the nearest wall-wall corner and down the corner to the baseboard, and finally over to a baseboard receptacle. Check local electrical codes first to be sure this method is permissible. It calls the least attention to the cord but usually requires more cord than is provided. Consequently, have an electrician rewire the track with the length needed.

See for Yourself: How Would It Look?

1. Tape a 15-ft (4.6-m) length of clothesline from the place on the ceiling where the track equipment will be to the baseboard receptacle.
2. Compare it with a longer length, taped to the ceiling-wall and wall-wall corners.
3. Decide which looks best.

Two sets of factors affect how the lighting appears from any low-voltage fixture. One set is the size of the lighted area, and the other is the intensity of the light.

Size of the Lighted Area

The size of the lighted area depends on the beam width, the fixture's distance from the painting and aiming angle, and the amount of illumination in the room. First, beams get wider as they get farther away from the light source. For example, a PAR narrow spot can be 1 ft (0.3 m) wide at 6 ft (1.8 m) away, and 2 ft (0.6 m) wide at 12 ft (3.6 m) away. Therefore, the closer to the painting, the smaller the lighted area is; the farther away, the larger.

The beam width originates at the light bulb. Both fixture and light bulb manufacturers publish catalogs with lighting performance data describing beam widths of bulbs at various angles.

Second, the fixture's distance from the painting and the aiming angle determine the size of the beam at the painting. These factors are interrelated. The aiming angle is measured in degrees from something. The aiming angle for a surface is 0° when the light comes perpendicular to that surface. Therefore, the aiming angle for a vertical surface (a wall-hung object or a display mannequin) would be 0° when aimed perpendicular to or directly to that surface. The aiming angle for a horizontal surface (the floor or a tabletop) would be 0° when aimed straight down. However, this definition is not consistent among manufacturers. Some consider 0° always as straight down, and others consider 0° as perpendicular to the object being lighted. Make sure which system is being used when consulting the lighting performance data in catalogs. Otherwise, what is 30° angle in one system is 60° in another; 45° is the same in both.

In this book, the aiming angle is considered 0° when it is perpendicular to the object being lighted. Therefore, the greater the angle, the closer the fixture will be to the painting and the smaller the distance from the wall. The smaller the angle, the farther the fixture from the painting and the greater the distance to the wall.

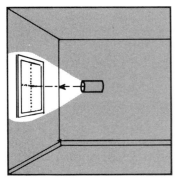

The closer to the painting, the smaller the lighted area.

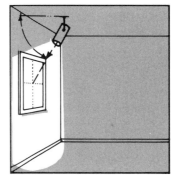

The farther away, the larger.

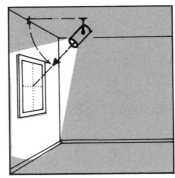

0 degree aiming angle.

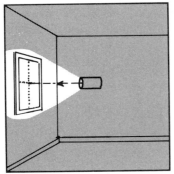

45 degree aiming angle.

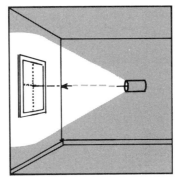

60 degree aiming angle.

Third, the size and amount of illumination in the room affect the area of light seen on the painting. If the wall is brightly lighted, the area of light on a painting will not be definable; it will be washed out by the other light. If the wall is dim, the area of light will be more distinguishable. Decide how dramatic the area of light is to be.

Rule of Thumb for Attracting Attention
In order to attract attention, the light on an object must be three times more intense than its surroundings; ten times the intensity will definitely be noticed.

Intensity of the Lighted Area

At any one aiming angle, the size of the area lighted is approximately the same for both the 25- and the 50-watt PAR bulbs, but the intensity of light is not. The difference in intensity is from two to five times as much. Of course, the intensity is also affected by the distance of the fixture from the painting. As the distance increases, the intensity decreases, because the further light travels, the more it diminishes. This phenomenon is called the Inverse Square Law. It makes the light from the sun tolerable; otherwise, the sunlight would burn us to cinders.

Likewise, the intensity seen is affected by the amount of light in the rest of the room and the color of the painting. The brighter the light

GUIDE TO LIGHTING EFFECTS

FACTORS	EFFECTS	
	Intensity of Light	Size Lighted
Very narrow spot bulb	brightest	smallest
Narrow spot bulb	to	to
Wide flood		
Very wide flood	least bright	largest
Close to painting	brighter	smaller
Far from painting	least bright	larger
Light direct on painting	brighter	smaller
Light at an angle	less bright	larger
Bright room lights	makes less visible	
Dim room lights	makes more visible	
Dark painting	makes less visible	
Pastel painting	makes more visible	

in the room, the less intense the light will appear on the painting. Conversely, the darker the room, the more intense the painting will appear. Dark paintings absorb more light than pastel paintings, thereby requiring more light to be illuminated. In spite of all the factors that affect the amount of light, just 50 footcandles will illuminate a painting sufficiently, in many cases. Yours, however, may be different.

The intensity of low-voltage light can be dimmed with an electronic low-voltage dimmer or a variable autotransformer.

lighting design

Most artwork should have the lighting aimed at a 45° to a 60° angle from perpendicular for the best effect. Glossy artwork (using shiny oils, acrylics, or glass frames) might require a steeper angle (60° to 70°) to avoid glare from the light. However, light at such an angle creates deep shadows and might emphasize too much texture in artwork. Check first to see if the emphasis is pleasing.

Lighting Just the Painting

When lighting just the painting, it is possible to superimpose the shape of the lighted area within the picture frame only, thereby making the painting appear to radiate light. It can be done with a low-voltage PAR bulb, because the shape of the lighted area is square to elliptical. A framing projector can do the same thing but requires

more watts to do it. (See Chapter 8.) To illuminate just the painting with any low-voltage bulb, determine the exact location for the fixture, because the distance determines the size (and intensity) of the lighted area. The best method to use is calculating carefully beforehand and making minor adjustments at installation time. The next best method is to see for yourself.

See for Yourself: What Is the Size of the Lighted Area?

1. If your fixture accepts a PAR bulb, have an electrician equip it with a 50-watt wide flood.
2. Wire the fixture by connecting it to an electric cord and plug with alligator clips, carefully wrapping the clips with electrician's tape and not touching them.
3. Hold the fixture at the ceiling where you could install it.

4. Move it forward or back to get the desired area of light.

5. If the area is too small with the wide flood, try the very wide flood. If it is too large, try the narrow spot or very narrow spot.

6. Make sure that glare is not going to be reflected from the artwork.

7. Mark the place on the ceiling where the fixture must be installed. (If a sofa or chair is in front of the painting, make sure that glare would not get into the eyes of a seated person. If so, install the fixture closer to the wall—at a steeper angle—and use a bulb with a wider beam.)

Light the Painting and the Wall

To illuminate your painting and the wall around it, locate the fixture so that the light comes from a 60° angle. For example, with a painting centered at 5 ft 3 in. (1.6 m) above the floor, install a track 1 ft 8 in. (0.51 m) away from the wall in 8-ft (2.4-m) ceiling heights; in 9-ft (2.7-m) ceilings, install the track 2 ft 3 in. (0.69 m) away; in 10-ft (3-m) ceilings, install 2 ft 9 in. (0.84 m) away. After installation, minor adjustments of up, down, left, or right can be made by moving the fixture itself. Track equipment is adjustable.

Sample Electric Cost

A low-voltage fixture consumes significantly less electricity than a regular line voltage fixture. For instance, using a 50-watt low-voltage bulb for 4 hours per night at 10¢ per kilowatt-hour, the electricity charge (including 20 percent more electricity for the transformer) is 72¢ for a month. By contrast, a standard voltage bulb, to do the same job, would need to be at least 75 watts and usually 150. As a result, a 150-watt bulb used for the same time period and at the same rate would cost $1.80. In addition, it would generate more heat, which might be uncomfortable during the hot weather or might damage the painting over a long period of time. Would you rather pay 72¢ or $1.80 per month to accomplish the same thing? A low-voltage fixture allows you to light up your painting and not your utility bill.

Dramatically light your painting within the frame only.

Light, less dramatically, both the wall and the painting.

putting your collection in the light

11

Collections are accumulated and displayed. But if they are not lighted, they do not get the attention they deserve. Whatever objects you collect—trophies, antiques, prints, sculpture, family photographs, or others—light can enhance their special characteristics.

what do you collect?

- Metal objects; light makes them shiny and glimmering.
- Glass; light is essential to give sparkle and brilliance.
- Ceramics; light reveals shape and subtle colors.
- Cloth, leather, or wood products; light points out texture and grain.
- Sculpture; light gives form, highlight, and shadow.
- Wall-hung art (photographs, prints, bas relief); light reveals colors and adds richness and authority otherwise not present without light.

In addition to complimenting a collection, the room itself benefits from light. It bounces into the room and softly balances the light from other sources. Light from a lighted collection can brighten difficult places, such as dark corners and dim sidewalls.

collections displayed on furniture

If furniture displays collections, purchase it already lighted or have it custom lighted. Already lighted furniture usually has incandescent light sources. However, consider replacing them with more energy-efficient fluorescent sources—new small screw-base bulbs, or tubes on a track or in miniature fluorescent holders. Fluorescent sources last longer than do incandescent sources and produce less heat that could damage collectibles.

Ready-made furniture such as china cabinets, breakfronts, secretaries, bookshelves,

Light reveals shape and subtle colors of ceramic collections.

Light a breakfront displaying Spodeware for
$1.24 a month at $.10 per kilowatt-hour.

étagères, and headboards are available with built-in lighting. Lighted furniture is excellent for both owners and renters. Furniture stores and interior designers feature them. However, ready-made furniture usually requires incandescent bulbs—sometimes tubulars (like the ones in showcases of department stores) and at other times reflectors (like a spotlight). Tubulars accent objects best from the front. Reflectors accent best from the top (downlighting) or from the bottom (uplighting).

The color of the incandescent light enhances the look of the wood in the cabinetry but detracts from the colors of blue and green objects. More important, heat is produced, which could fade colors or damage objects, particularly paper, leather, or other nondurable goods.

As energy gets scarcer, furniture manufacturers will offer cabinetry with more efficient light sources. In the meantime, if cabinets are chosen with tubular incandescent lights, seek a reflectorized bulb, which spreads the light. They are hard to find but worth the effort. If it is unobtainable, back up a tubular bulb with metal, spreading the light and protecting the cabinetry from the heat. Likewise, if the cabinet has a reflector bulb (flood or spot), choose glass shelves to allow the light to go all the way through; otherwise the watts are wasted on just one shelf.

Whereas a 50-watt incandescent will illuminate a 12-in. (30-cm) shelf, a 15-watt fluorescent will illuminate a 33-in. (84-cm) shelf for 35 watts less. Watts are saved, and the lighted area is larger. Fluorescents are not often incorporated into already lighted furniture, but they can be installed by a cabinetmaker or a careful electrician. Fluorescent tubes, because of their straight-line shape, fit well into cabinetry. Two kinds are available: a tube on a miniature track or a midget strip. A 15-watt tube can be mounted on the ready-made miniature track and positioned under a glass or wood shelf. It is out of sight and hooks up to the electricity with a cord and plug. In addition, the miniature track accepts an incandescent low-wattage nightlight bulb or a small spotlight. Other fluorescent fixtures—midget strips—for small fluorescent tubes are less than 24 in. (61 cm) long, ⅝ in. (1.6 cm) wide, and 2½ in. (6.4 cm) deep. They are called miniature bipins. Purchase the size that fits the shelf.

- For backlighting, tuck them neatly behind or below the shelves.
- To accent, place them on the left or right, inside the cabinet door.
- To uplight, secure below a frosted glass shelf, to hurl their light up.
- For frontlighting, slip them beneath the shelf above, at the front.

Whatever the position, hide the fixture behind a structural portion of the cabinet or a piece of wood trim added on.

Installation and Effects

Different materials are enhanced by light from different locations. Metal objects—guns, trophies, sculpture—respond better to light from the sides to accentuate the surface of the metal and enhance the shine. Glass objects—bottles, boxes, ornaments—respond better to light from below or above. The edges of the glass will take on the light and glow brilliantly. Glass objects on lighted frosted glass glow as though they were bursting with luminescence. Ceramics, cloth, and leather respond better to light from the front, above or on the sides, revealing the shape, color, and texture. If the color is not important, backlighting reveals the silhouette and calls attention to the shape of the object.

Sample Electric Cost
The light for a breakfront with three shelves consists of six fluorescent ready-made fixtures of 15 watts each, totaling 103 watts, including 15 percent more electricity for the ballast (the built-in regulator). The light for 4 hours per night at 10¢ per kilowatt-hour costs $1.24 per month.

---

free-standing collections

Free-standing collections are three-dimensional objects that stand alone, such as sculpture, large antique vases, and oriental screens. Free-standing objects have unique forms, solid or open, which require highlight and shadows to reveal them. Light creates highlights as well as shadows. Without light, objects can appear flat and uninteresting. Only direct light gives birth to highlights and shadows—and it must be powerful. It can also be energy-efficient if it comes

Freestanding objects demand light.

Installation

The light from the front and sides can originate at a ceiling with a low-voltage fixture. The light from behind can also; or else it can be an uplight with a standard-voltage fixture, provided it does not glare in anyone's eyes.

A ceiling fixture can be recessed, surface-mounted, or on a track or canopy. The track or canopy can be installed either permanently into a ceiling junction box or temporarily hooked up by a cord and plug to a baseboard. A surface-mounted or recessed fixture, on the other hand, must be permanent. Owners can permanently or temporarily install fixtures or tracks, and renters can install tracks temporarily. A floor uplight can sit on a weighted base and can be hooked up by a cord and plug. It is a temporary installation for both owners and renters.

Fixtures

Low-voltage fixtures are adjustable. A transformer (an apparatus to change the voltage of electricity) is built into the fixture and reduces the standard current (120 volts) to 12 or 5.5 volts. The fixtures are manufactured in compact shapes, and they utilize low-voltage bulbs. They make it possible to capture the best highlights and shadows, which is very dramatic and artful.

Bulbs

The bulbs are either a small multifaceted reflector bulb (MR), similar to a projection bulb, or a PAR bulb, similar to an automobile headlight. These bulbs put out very controlled beams of light in widths that range from extremely narrow spot to very wide flood. They supply a variety of beam areas and brightnesses, and they are available in 20 to 75 watts. (See Chapter 10 for more low-voltage details.)

from low-voltage fixtures and bulbs. With low-voltage fixtures, heat is also low and the bulbs last longer.

Each free-standing collectible must be tested with light to determine the best way to create the highlights and shadows. One light is unlikely to make an object appear dynamic; two or three are frequently needed.

See for Yourself: What Will Give the Best Effect?

1. Use a flashlight, an auto trouble light, and any portable lamps you have.
2. Shine the brightest on the front from above, one from the side, and another from behind.
3. Move them until you find the positions that create the greatest visual impact.

Not all of the light should be equal. Some lights should be brighter, some softer. Consequently, after installing the fixtures, try several wattages, using the least first, and working up to a higher wattage if necessary.

Sample Electric Cost

A sculpture highlighted by two multifaceted reflector 50-watt low-voltage bulbs for 4 hours each night costs $1.44 per month, at 10¢ per kilowatt-hour, including 20 percent wattage for the transformer. Low-voltage fixtures cost one-third more to purchase than do regular voltage fixtures, but they consume two-thirds less electricity over the long haul.

---

collections hung on the wall

Collections that hang on the wall include photographs, prints, paintings, tapestries, macramé, and bas-relief. How to light them depends upon how serious the collector is, where the collection hangs, how often it is changed, and whether the collector is an owner or a renter. A serious collector will be willing to spend more to enhance his or her collection. A casual collector will want enhancement too, but at a smaller cost. A permanent collection on one wall can be lighted with wall-to-wall permanent lighting. A changeable collection requires changeable lighting. Owners are able to build in lighting. Renters must use removable lighting.

Owners with Serious Collections on One Wall

Owners with a serious collection hung on one wall will want to illuminate the whole wall with energy-efficient fluorescent wall-washing. It will flatter the collection, the room, and the occupants, particularly when equipped with deluxe or prime-color fluorescent tubes. (See built-in fluorescent cornice light in Chapter 5.)

The collection can be changed without disadvantaging the lighting effect. If the collection happens to be framed with glass, fluorescent light illuminates it glare-free. If the collection is removed completely, the lighted wall will continue to flatter.

Owners with Serious Collections on Several Walls

Owners with a serious collection spread out on many walls should check Chapter 10 for various methods of lighting one piece of wall-hung art. Whatever method is chosen, select the best items in the collection to be illuminated and also decide which rooms could benefit from the extra light. In most cases, extra light helps semipublic rooms: in residences—living rooms, family rooms, dining rooms, and hallways; in businesses—waiting rooms and executive offices.

Some owners may want to invest in a low-voltage track canopy and track fixture. The track fixture needs to be adjustable but does not need to slide. The track fixture could be detached from the canopy and put away if the collectible is removed. Purchase a low-voltage track fixture and a track canopy for each collectible to be lighted. Equip the fixture with a low-voltage bulb in the beam width that matches the size of the wall hanging. Try the 20-watt PAR flood first if the wall hanging is not large. Otherwise, try the 50-watt very wide PAR flood. No rule of thumb is available for lighting collections, because so many variables need to be considered—color of the object, distance from the object, and the total light in the room, to mention a few. (See Chapter 10.)

INSTALLATION AND EFFECTS

The size and brightness of the area lighted depends first on the type of bulb chosen and second how far away it is. The closer the bulb, the smaller and brighter the lighted area is. The further away, the bigger and softer it is. The bulb will produce effects ranging from a bright rectangle of light to a soft, pool-shaped glow.

An electrician should pretest the track canopy before deciding where to install it. Typically, it should be installed about 3 ft (0.9 m) out from the wall and centered on the collectible. Make final improvements after the fixture is clipped in to the canopy. Test the three wattages and three bulb types available. One will satisfy.

If the collection is framed with any kind of glass, avoid creating a glare. Glare obscures—especially on photographs where the details are small. Glare can be avoided by installing the fixture as close to the wall as possible and grazing the picture with light.

Wall-hung objects can be lighted by a cornice light.

Owners as Casual Collectors

Owners, with a casual collection, either changing or static, should use a fluorescent wall bracket. Mount it high at the top of the wall or just above the wall object. Brackets effectively light wall objects up to 8 ft (2.4 m) wide or a group of smaller objects. The bracket must be larger than the object, but not much larger. Brackets must conform to the standard increments dictated by the sizes of fluorescent tubes: 4, 6, or 8 ft (1.2, 1.8, 2.4 m). Ready-made wall brackets are available, in 4-ft (1.2-m) and 8-ft (2.4-m) sizes. They are either box-shaped or tube-shaped. They are available in many colors. They mount flat to the wall, or hang by a stem from the wall or ceiling. On long walls they can hang in tandem, stretching in modules divisible by their length.

Likewise, fluorescent wall brackets can be custom-made. They are usually box-shaped and built like a dropped soffit, but narrower, using only one bulb. Therefore, a bracket is built approximately 6 in. (15 cm) wide. (Follow the directions in the dropped soffit section in Chapter 9.) Custom-made brackets can be positioned at the top of the wall or directly above the object. Because the 4-ft (1.2-m) fluorescent bulb is the easiest to obtain and the quietest with a good-quality ballast, build your bracket to use this size, if possible. If a 4-ft (1.2-m) size is too small and two 4-ft (1.2-m) fixtures are too big, try 3- or 6-ft (0.9- or 1.8-m) sizes.

INSTALLATION

Center the bracket over the object. Avoid a cord and plug connection, if possible. Install the bracket through a junction box in the wall or ceiling, provided that electric wires are accessible. Owners can specify a junction box when developing the building plans, or when relighting they can have an electrician gain access in an attic or pull wires from elsewhere.

Renters with Serious Collections on One Wall

Renters with a serious collection on one wall can light it with a track and several low-voltage fixtures. The track can be electrified by a cord and plug and can be mounted on the ceiling for the duration of the lease. At moving time, it can be moved to illuminate another wall in another location.

FIXTURES AND BULBS

Select only low-voltage bulbs in the sizes that produce the amount of light you want. It is costly to dim low voltage. It must be done through the primary electrical source to the transformer and requires a special low-voltage dimmer.

In making the bulb selection, decide if an even wall-to-wall wash of light or variable pools of light is wanted. For even light, space the fixtures equally and equip them all with the same size bulbs. For variable light, space the fixtures as dictated by the position of the objects on the wall. Equip the fixtures with either the same bulbs for equally bright pools of light or different bulbs for brighter and softer pools of light. A mixture of spots, floods, or very wide floods produces different amounts of light for a variably lighted wall. (For bulb sizes and brightness amounts, see Chapter 10.)

INSTALLATION AND EFFECTS

Check local electrical codes to determine if a cord and plug are permissible. If so, position the track about 3 ft (0.9 m) from the wall, running the cord along all the corners—ceiling, wall, and baseboard. This installation makes a temporary connection look permanent; it is the least obtrusive way. Purchase enough cord to reach the baseboard receptacle by way of taking the cord to the closest wall, moving along the ceiling-wall corner, to the wall-wall corner, then down to the baseboard and over to the receptacle. The added expense and effort are well worth the added results. Finally, adjust the fixtures to illuminate either evenly or variably. Even lighting creates a broad wash of light at the intensity deliverable by the wattage chosen. Variable lighting emphasizes lighted objects, not the wall. Either type of lighting is appropriate; both are spectacular.

Renters with Collections on Several Walls

Renters with either serious or casual collections on several walls should use portable lamps—floor lamps, table spots, picture lights, or possibly pin-ups.

FLOOR LAMPS

Floor lamps sit on the floor, reaching out to highlight a nearby wall object. Some floor lamps have multiple arms to spot several objects at once. Some have multiple lights but not arms, to spot not quite as far. Some are slim shafts of fluorescent light. If positioned in pairs, they can illuminate wall objects from two sides.

TABLE SPOTS

Tabletop spots sit on top of furniture or clip on a shelf, aiming at a wall object. Tabletop spots are scarcer than floor lamps and are mostly slick contemporary shapes; some blend with traditional furnishings. They accept standard or low-voltage bulbs. The type needed is determined by the distance away from the object and the brightness desired. The longer the distance, the higher the wattage required. To be energy-effective, place the spot lamp close to the object on shelves or tabletops next to the wall. In most situations, a spot lamp with a 50-watt standard voltage reflector bulb does the job.

PICTURE LIGHTS

Picture lights illuminate pictures by hanging on the wall or on the frame. They use tubular incandescent bulbs. The tubular R-type bulb is the best. It has a silver reflector strip that allows all the light to be directed to the picture. Picture lights are available in a variety of sizes, most often 10, 18, and 28 in. (25, 46, 71 cm) long. Likewise, pictures can be lighted on an easel away from the wall. Either way, the power of picture lights is limited. They throw noticeable light down only 3 ft (0.9 m). Therefore, plan to light a picture that is no more than 3 ft (.9 m) high. Further, choose a light that is at least two-thirds as wide as the picture. For example, use a 10-in. (25-cm) picture light for a 16- by 36-in. (41- by 91-cm) picture. When deciding to buy, remember that within the limits of its candlepower (the ability to produce light) the farther out the picture light extends from the picture, the farther it throws the light down.

TRACK PICTURE LIGHTS

Some track lighting fixtures can light pictures.

Tabletop spots.

Use one or two 25-watt reflector incandescent bulbs. The track can be mounted above the picture on the ceiling or on the wall. Adjust the fixture and aim the bulbs to provide the best light.

Sample Electric Cost

For 4 hours a night at 10¢ per kilowatt-hour, these two 25-watt bulbs consume 60¢ worth of electricity per month to illuminate one picture in a collection.

PIN-UPS

A pin-up (a cord-and-plug connected, wall-hung fixture) can be a picture light in some cases. A pin-up is not always aesthetically successful as a picture light, because it is a large and obvious gadget. However, pretest it in your space and maybe it can be arranged to be artistic and harmonious.

At most, picture lights throw light down only three feet.

read in bed
but
do not disturb

12

Commands of "turn out that light" and retorts of "I only have one more page" echo in many bedrooms, because some people enjoy reading in bed and others want the lights turned off. This conflict is a lighting problem that most of my lighting clients ask me to solve.

Light for bedtime reading usually comes from lamps on bedside tables. But bedside lamps do not illuminate reading material well enough and definitely do not confine the light enough to permit one person to sleep while the other reads. Further, bedside lamps are not usu-ally adjustable and cannot be moved to permit the most comfortable reading position. Like-wise, in new construction, room sizes are often too small to accommodate bedside furniture with lamps. This is no great loss. Wall fixtures are actually better. Adjustable wall fixtures per-mit one person to read while the other sleeps, if they have opaque shades (shades that do not transmit light). Use a single, a double, or two single fixtures. Install them at the head of the bed so the light can be directed away from the sleeping partner.

installation

Wall fixtures are either wired directly through the wall or plugged into a baseboard receptacle. The through-the-wall method requires an elec-trical junction box. This method is called either wall mount or outlet-box mount. (Manufactur-ers' terminology differs.) The plugged-into-the-wall method requires a cord and plug, and it is called either pin-up or cord and plug. When or-dering the fixtures, specify which type you want. Owners can take advantage of outlet-box mount; renters must resort to the cord and plug. An outlet-box mount should be incorporated when building and can be incorporated when relighting. When building, owners should spec-ify an appropriately placed junction box before the building process begins. When relighting, have an electrician pull wires from a baseboard receptacle to electrify a junction box. In most cases, an electrician is required to install an out-let-box mount fixture. Nonetheless, it is the slickest method, because wires are not visible.

If outlet-box mounting is not possible, pur-chase a fixture with a cord and plug. The cord usually hangs below the fixture, unless you are clever about hiding it or unless you choose a fixture that has a tube to hide part of the cord—

One person can read and one can fall asleep
with a wall-mounted fixture.

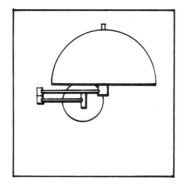

Wall-mounted fixture.

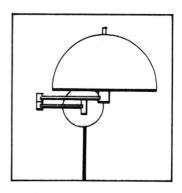

Pin-up fixture.

appropriate on some styles, inappropriate on others. No electrician is needed for this installation, and renters find this method the best.

Rule of Thumb for Mounting a Reading Light
Mount the wall fixture 30 in. (76 cm) above the mattress.

Modify this rule if necessary to make sure the fixture is within easy reach from a semireclining position, in order to be able to adjust it or turn it on and off. In addition, the rule may need to be modified according to the headboard height. Center the fixture above the bed, and point the light to the reader's side. Finally, for the greatest convenience, locate the on-off switch either on the fixture or on the wall plate behind the fixture.

fixtures

Choose a fixture with a shade that does not allow light to escape—usually metal. However, metal absorbs heat from the bulb. Some fixtures are designed to dissipate part of that heat through perforated shades, letting light escape also. To confine the light, choose only solid shades or high-quality fixtures that dissipate the heat and not the light.

Select an adjustable fixture—with an adjustable shade, an adjustable arm, or both. An adjustable shade swivels and allows the light to be directed exactly where needed, but be careful to choose one that does not droop. On the other hand, an adjustable arm, hinged or folding, allows the fixture to reach out farther. Some arms reach as far as 34 in. (86 cm).

Choose a fixture style that is in concert with the room. Often, manufacturers make the same style in both outlet-box-mount and cord-and-plug types. Also, they sometimes make a single or a double version in the same style.

In addition, two track lights can be used for reading lights. Track fixtures can be mounted together on an electric track or singly on a track canopy. Both are adjustable. Track fixtures are movable left to right; canopy fixtures are not. Like other wall reading lights, the switch needs to be on or near the fixture.

Tracks are available in 2- or 4-ft (.6- or 1.2-m) lengths and can be cut to match the width of the bed. Even so, electrical tracks may present a hazard. Although the electrical wiring is well recessed, a paper clip or other small metal object can be poked into an open track and instantly conduct the electricity to the person holding it. Therefore, track lights are not a good choice for areas where children and others not thoroughly responsible for their actions are present.

The electrical connection for a track is made through a junction box or by a cord and plug. Most track canopies require a junction box, but some can be ordered with a cord and plug.

light source

All of the fixtures described here are designed to utilize a screw-base bulb. Tubular, standard A, and reflector incandescent bulbs are available, and a fluorescent screw-base bulb is also available. Manufacturers specify the type and size bulb to use with each fixture, and they often stamp it on the inside of the shade. However, lower wattages may suit your individual needs and save electricity in the bargain. In general, enough light for short-term reading can be ex-pected from a 50-watt reflector bulb. But if the intention is to read for several hours, use a 75-watt incandescent bulb (or its equivalent in fluorescent) and another light on in the room near the bed, providing well-balanced light. Do not attempt to read for a long time without additional light. Eyes fatigue from adjusting to the bright light on the book and the dull light in the room.

hobbies
demand light
13

Hobbies give us a sense of satisfaction and joy. Many are visually demanding. They require sufficient illumination to be accomplished easily. Without light, hobbies become tedious and slow. Adjustable fixtures offer the most options for getting light to a hobby, but do not rely on adjustable fixtures alone; others are needed elsewhere in the space.

needlepoint or other handwork

More light is needed for needlepoint and other handwork than for reading or watching television. More often than not, handwork takes place in the living room or other family relaxation space. Usually, standard light sources—portable lamps and sometimes ceiling fixtures—are the only sources available. But an adjustable lamp that puts out additional light is also needed. If the eyes are not too good or the handwork has little contrast and small details, such as 16-point needlepoint canvas, a 200- to 300-watt standard voltage or 25-watt low-voltage incandescent bulb should be in the lamp. Make sure the lamp has the capacity for such a bulb.

The lamp can be wall-mounted or placed on a tabletop or the floor.

An excellent portable lamp for hobbies with small details is a fluorescent circle lamp with a magnifying glass in the center of the shade. It clamps or sits on a table edge and adjusts to any position. These lamps are excellent gifts for people who enjoy visually demanding hobbies, particularly for those who cannot see as well as they used to. Whichever type of adjustable lamp you choose, always have at least one other light source on in the room to provide background lighting. Otherwise, the eyes can become strained.

woodwork, metalwork, and other workbench crafts

Workbench hobbies may be visually difficult or may be simple. Difficult ones require more illumination. If the contrast of materials used in the hobby is great, the details are large, the length of time is not too long, and the eyes are good, the amount of illumination can be 20 to 50 footcandles. As the contrast lessens, the details get smaller, the time lengthens, and the eyes worsen, the amount of light required goes up. It can go up to 200 footcandles for metal engraving, gem polishing, and jewelry making.

For both day and night, the lighting should be balanced but not bright. The light from a bright sunny window, for instance, should be reduced to balance the amount at the workbench. The workbench fixture should be out of the way—that is, ceiling-, wall-, or shelf-mounted, or suspended. A fluorescent fixture is the most suitable. Position it so that the tube does not reflect an image on any shiny surface—over the front edge or on the two sides of the bench. If mounted over the front edge, the fixture should be about two-thirds the length of the workbench. For example, a 6-ft (1.8-m) workbench should have a 4-ft (1.2-m) fluorescent fixture. Fixtures with reflectors send out more light than those that do not. Ceiling fixtures require a wall switch; suspended, shelf-mounted, and wall-mounted ones do not. Fluorescent fixtures can hold one, two, or four tubes. Choose according to how much light is needed and how much is absorbed by the interior colors. Normally, two tubes are needed, but four may be required for hobbies that are difficult to see. Further, if matching colors is important to the

Workbench fixture should have a reflector.

hobby, special color-matching tubes are available; otherwise, the common cool-white fluorescent is satisfactory. (See Chapter 20 for color-matching details.)

Owners will want to make their fixture installation permanent. Renters can use several inexpensive fluorescent strips with the convenience of a switch on the cord with a plug.

painting, sketching, or collage

If someone paints, sketches, or does collage, colors are critical and the illumination must reveal the correct colors. The amount of light must be sufficient, and the source must contain all the colors in the spectrum. For this reason, artists paint by daylight at a north window. Daylight contains all the colors in the spectrum. The north daylight is not direct or harsh, yet it is sufficient. Not all hobbyists have a north window, and some must paint at night. Therefore,

lighting fixtures must provide the required illumination.

The amount required is as much as in a well-lighted office. The fixtures must light the space uniformly because the hobbyist often copies something elsewhere in the room, and the objects to be copied must also be well lighted. Therefore, ceiling fixtures are suitable; they can be recessed, surface-mounted, or suspended. In addition, for seeing fine details, a light fixture—

pendant or portable—can be located at the easel or the work surface itself.

Should the light source be incandescent or fluorescent? If the end product of the hobby is in color and you know under which light source it will be seen, use the same source. Otherwise, choose the source that works best for the space.

Fluorescent sources can render colors correctly. The color-matching types are used in printing shops and hospitals, where colors are critical. They are more expensive, but like other fluorescent tubes they last a long time and use little energy. (See Chapter 20.)

sewing

Using a sewing machine varies in visual complexity with each sewing job. Many components vary: thread and stitch size, contrast of thread and fabric, and reflectances of fabrics. Maximum light should be at the sewing-machine needle. For complex jobs, 200 footcandles are needed, and the rest of the tabletop should be illuminated to 20 footcandles.

Additional fixtures for additional light need to be located on a wall, shelf, or ceiling, or they can be free-standing. Their position must not create a shadow from the user's hand.

playing a musical instrument

Musical scores vary from simple with large notes and dark lines, to complex with small notes and many line notations. Casual musicians are not as concerned with speed and accuracy in reading a music score as advanced or professional musicians. Casual musicians with simple scores require normal reading light (approximately 20 to 50 footcandles); professionals with complex scores can require as much light as do drafters (approximately 100 to 200 footcandles).

Renters should be aware that portable lamps that mount on the music rack do not illuminate the full page well and should be used only when no other solution is possible. Floor-mounted portable lamps with several light sources can illuminate the score and the general surroundings better.

Owners should be aware that built-in fixtures should be above the musician's head. If the fixture takes an incandescent bulb, it should be adjustable and strike the music rack between a 30°- to 60°-angle from behind the head or on either side of the music. If the fixture is fluorescent, it must not create glare or reflections on the scores.

other hobbies

Follow the illumination requirements similar to those already described in this chapter. If the detail is large, the illumination requirements are low; if they are small, the requirements are high. If the hobby is visually complex, the requirements are high; if it is simple, the requirements are low. For example, jewelry making—very visually complex—requires more light than pottery making—relatively simple. In addition, if 8 hours a day are spent on a hobby, more light is needed than if half an hour were spent. More light is needed for people over 40 years old. Further, if the room surfaces are dark, more illumination is required than if they are pale. Determine whether the maximum or minimum illumination is needed by the requirements of the hobby. Pretest whenever possible and be willing to add more light for your hobbie's requirements when it is necessary.

light for
older eyes

14

As we grow older, our eyes need more light in order to see. After age 55 the eyes do not see as well, sometimes even in good light. Young people think they can see in any light; they often study in a dim room without turning on the available lamp, which is often right next to them. On the other hand, older people know that light acts as a magnifier—they take the needle to the window to thread it or the phone book to the lamp. Older people (particularly those older than 65) should have colorful, detailed, and correct visual information, which can be provided by good light. Some older people claim that they do not need light and that they must save money. Scrimping on light is a false economy. Injury or depression costs much more.

where is light needed?

Ideally, light is needed everywhere. At a minimum, it is needed in the room with the television, at a mirror, at kitchen counters, at stairs, near bathtubs or showers, and in closets. When the television is on, one light should be near but not on top of the television set and another one elsewhere in the room. Television should never be viewed in the dark; it is too tiring for the eyes. Further, light is needed at the bathroom mirror, positioned along the sides to illuminate the whole face. Most bathroom lights are behind the person standing at the mirror, illuminating the top of the head and leaving the face in ghoulish shadow, which is not flattering to anyone.

At kitchen counters, where sharp tools are used, utensils are identified as dirty or clean, and wholesomeness of food is differentiated, light is needed. Yet most kitchens are lighted from the center of the room. This position puts people between the light and what they are doing, causing deep shadows. This is not very safe for anyone at any age.

Light over tubs or showers permits personal care and helps ease getting in and out. Light in closets helps with searching and identifying. Finally, light is needed at steps and other hazardous places, because a potential outcome—personal injury from falls—is so devastating. Many times, older people intend to save money by turning off the light and walking up the stairs in the dark. However, the cost for an injury is infinitely higher than the pennies saved on an electricity bill. After all, a 100-watt incandescent bulb can burn for 1 hour at 10¢ per kilowatt-hour for 1¢. Older people are penny wise but dollar foolish if they think that keeping themselves in the dark will greatly help their budget problems. More money would be saved by heating water less often, because hot water heated electrically averages about 30 percent of the bill and lighting averages about 5 percent. Without adequate light in living spaces, older people are buying into a duller and drabber existence. Light can establish a cheerful environment and can make objects visible for a safe, pleasant old age.

Because the ability to see details has decreased, older eyes need greater quantities of light—about twice as much as young people. Because sensitivity to glare has increased, older eyes need glare-free direct and reflected light. Glare can disable. Because most older folks wear glasses—almost 100 percent over age 65 do—light should not create false illusions of depth. Reflections from a glossy surface can confuse and create illusion. Because contrast sensitivity has declined, older eyes need a suitable relationship of light nearby to light farther away. Contrasting areas of light can cause difficulties in seeing. Because light draws attention, it should focus and guide older people through architectural changes, especially potentially dangerous ones. Because light fixtures need maintenance, fixtures for older people who maintain their own residences should be ser-

Changing bulbs should not require a ladder.

viceable from the floor. Changing bulbs should not require a ladder.

what kind of light should be supplied?

In general, older folks should use soft, diffused light that minimizes shadows, with at least two light sources in each space. Light sources should be large as possible but not bright, such as big lampshades with low luminescence, large opal diffusers, or the largest fluorescent ceiling fixture altered to use fewer tubes than intended. Whenever fluorescent light sources are used, choose warm-white deluxe or prime color tubes. They are preferred by most older people. Whenever incandescent light sources are used, choose shades for lamps that are white or off-white and allow the light to come through. Colored shades impart color to the light.

Interior spaces should be equal in brightness; a bright room should not lead into a dull room. Light should be positioned to guide and call attention to important facilities in a room. No bare light bulbs should be in view. Shield all bulbs with shades or diffusers, especially when seen from above, such as when you are descend-ing stairs. Use coated incandescent standard A bulbs, not frosted or clear ones. Frosted and clear bulbs have a hot spot; coated bulbs do not. Use strong directional light—spot or flood bulb, accent or track fixtures—only to illuminate sewing or other small details for hobbies and handwork. Do not use a brightly lighted work area in a dark room; light up the room, too.

Do not place electric cords where they can be tripped over. Use tabletop dimmers so that control is at the fingertips. Put crossbars or other visual identifications on all large glass doors, indicating their existence. Light will reveal the identification and prevent many bumped noses. Do not use open ceiling-mounted downlights, especially over hard or glossy surfaces—vinyl floors, furniture of polished wood, or glasstopped tables. Use fluorescent light sources in ceiling lights; incandescent bulbs need replacing too often.

how can light be supplied?

- Pendants with pulleys, which permit easy replacement of light bulbs.
- Shaded chandeliers with downlights.
- Inexpensive fluorescent strips mounted as their own shield under kitchen upper cabinets.
- Shielded wall-mounted fixtures for general room illumination.
- An oversized but underwatted ceiling fluorescent fixture.
- A whole wall washed with light in the main living area.
- As many lamps around the room as possible controlled by one switch.
- A wall bracket at either side of the bathroom mirror.
- Fluorescent chain-hung pendants.
- A fluorescent light over or an incandescent light near the tub or shower.
- An incandescent ceiling light, if it is the only choice, dimmed continually to lengthen bulb life and reduce replacements.
- For general illumination, not reading, lamps equipped with fluorescent circular tubes of 44 watts that give more light than a 75-watt incandescent or with a fluorescent screw-based bulb that give more than a 60-watt incandescent.
- A special clamp-on magnifying lamp, particularly good for small details.
- The bottom of every lampshade at eye level.
- Three-way incandescent bulbs of 50, 200, and 250 watts, permitting different quantities at different times, in lamps with a diffuser under the shade.

how much light must be supplied?

Because spaces differ in size, amount of daylight available, and interior finishes, and because fixtures differ, it is impossible to give advice regarding particular wattages. (To measure footcandles, see Chapter 2.) Generally, the Illuminating Engineering Society advises that the amount of light required to see detail and color over age 55 is at least one and a half times the amount needed for people up to 40 years of age.

- Eating dinner requires not 10 but 15 footcandles.
- Reading, ironing, desk work, laundry, cooking at the stove, and looking in a mirror requires not 20 but 30 footcandles.
- Working at the kitchen counter and painting at an easel require 75 footcandles.
- Playing the piano requires no less than 30 and possibly as high as 200 footcandles, depending upon the size and contrast of the music score.
- Hand or machine sewing and workbench tasks require up to 200 footcandles.

how much does it cost?

These lighting objectives can be accomplished with reasonably priced fixtures and lamps because the degree of improvement in lighting does not have to be related to the amount of money spent on it. Therefore, look for suitable fixtures during lighting showroom sales. Or check the prices at large department stores, especially those with catalog sales. On the other hand, most any time a throwaway fluorescent strip can be purchased for less than $10.00.

If there is any time in one's life to pour on the light, it is when one is older. The investment in lighting overall is very cheap; it pays back in living quality for the present occupant and sells well to the next. Ask any real estate agent what is the first thing he or she does when showing a residence to a prospective buyer. They turn on the lights. People are sold on well-lighted, cheerful environments. Why not make the environment cheerful for yourself or someone you know and love?

outdoor and indoor light for reassurance

15

Light outside and inside can be reassuring; it does not have to be bright, just well distributed. Outside, the larger the lighted area, the greater the surveillance potential. Therefore, illuminate large areas outside windows. In addition, illuminate all pathways—sidewalks, steps, doorways, and driveways—by pointing the light at the pathway, not at the person who might be there. Light in people's eyes blinds and hinders rather than helps them. Well-positioned outdoor light can create a pleasing, comfortable, and helpful environment.

Inside light keeps away whatever comes out in the dark and allows people to move around safely. It acts as a greeting in empty rooms and reassures with a warm glow, whether as a single nightlight in a wall receptacle or as another low-energy light source.

reassurances outside

From the inside at night, uncovered windows of any size become mirrors reflecting images and revealing only blackness beyond. In low-rise structures, the blackness can become a problem. Owners can permanently install outside floodlighting. Renter, on the other hand, must rely on temporary floodlighting. In high-rise structures, the blackness beyond the windows on high floors is the sky, not potential prowlers.

Floodlights should be shielded or protected from the weather by roof overhangs or by the fixture itself. A shielded fixture, sometimes called a bullet, needs to be 9 in. (23 cm) deep. Mount it on the structure, in the trees, or on 10- to 20-ft (3- to 6-m) high poles.

Shielded flood bulb.

Conceal floodlights in the trees.

Illuminate pleasing objects.

Illuminate pleasing objects.

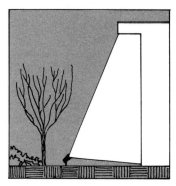

Wash the building.

Rule of Thumb for Floodlights

Floodlights will illuminate an area as large as one to two times the distance of the fixture from the ground.

Consequently, if the structure is two stories high and the floodlight is at the roofline, the light is comfortable and glare-free, projecting 17 to 34 ft (5.2 to 10.4 m) out. If the structure is one story, the light is not as glare-free and projects only 8 to 16 ft (2.4 to 4.8 m). It can be in the line of sight. Never point floodlights at driveways, walks, or entrances, as they would obscure people's vision. Likewise, do not floodlight objects that should be ignored. Illuminate objects that are pleasant to look at from the inside—trees, shrubs, walls, and fences. Lighting the outdoors beyond the windows visually expands the indoor space and makes it appear larger.

Choices for outdoor floodlights are regular-voltage incandescent PAR bulbs (65 to 150 watts), deluxe mercury (40 or 50 watts), or low-voltage PAR bulbs (25 to 100 watts).

On commercial structures, permanent outdoor floodlighting should wash the building's facade. It discourages vandalism and break-ins but also assists surveillance at night from law enforcement and security personnel. Further, light on the exterior of any building, whether commercial or residential, can create a favorable impression on passers-by.

Wash the facade of a commercial building with floodlights with one of these three ways:

- Floodlights on the building roof lighting down and close into the base of the structure and not aimed out more than a distance of one to two times the height of the building.
- Floodlights on the ground located a good distance away and hidden.
- Floodlights on poles some distance away and washing the facade.

In all cases, use wide-beam floodlights. The choices are incandescent or high-intensity discharge (sodium, mercury, and metal-halide) bulbs. Incandescent is the cheapest to install but the most expensive to operate. High-intensity discharge is more expensive to install but the cheapest to operate. Sodium gives a golden color, which may or may not enhance the color of the structure. Check it out first. Metal-halide and deluxe mercury produce reasonable color, which makes objects appear normal—reds look red enough.

Wattages from 50 to 1,500 are available to

All drawings on this page are reprinted by permission of Kim Lighting.

illuminate in any surrounding, whether dark or bright.

All commercial outdoor lighting fixtures should be indestructable, because the usual break-in method is to eliminate the light source and to enter the building in darkness. Likewise, light fixtures are sometimes beacons for pranksters to break, just for kicks.

Lighting for Entrances and Exits

Light all doorways for safe passage. Several kinds of outdoor equipment can be used:

- Wall fixtures mounted at eye level, about 5 ft 6 in. (1.7 m) high, in pairs (one on either side for major entrances) or singly (on the same side of the door as the lock for other entrances).
- Wall lanterns with side brackets (overportal lanterns) above the main entrance.
- Ceiling fixtures (surface-mounted or suspended) on the underside of a flat roof overhang on one-story structures. Make sure there is enough clearance to swing the door.
- Downlights recessed above the entrance, if the roof overhang is flat and enough ceiling depth is available. A round fixture illuminates a 1- to 2-ft (0.3- to 0.6-m) smaller area than a square or rectangular fixture. If a recessed fixture is chosen for the entrance, it can also be used to illuminate a solid exterior wall. The light calls attention to the wall surface, adding emphasis. Space the fixtures about 10 ft (3 m) apart. Equip them with 34- to 60-watt standard A incandescent bulbs, unless more is needed for impact. Do not use recessed fixtures over windows in the roof overhang; the light shines inside and might not be welcome.

People walking toward an entrance focus on it. Consequently, any glare is obnoxious. Glare is essentially uncomfortably bright light. It is created by bright direct light from the fixture and dim surrounding light. Some fixtures prevent glare by shielding and redirecting the light up and down. Some prevent glare by concealing the light with a diffuser. Some fixtures are transparent and allow the bulb to be seen. They are beautiful. Do not spoil their beauty by mak-

ing them glaring. Use only clear bulbs and make the surroundings well illuminated. Under such conditions, a clear bulb, in 25 watts or less, should not offend. Frosted or coated bulbs are not aesthetically pleasing in transparent fixtures and do not reduce the glare.

Some transparent fixtures contain downlights, adding a brighter light below. However, downlights do not give the authority to use a higher wattage above. One of the most successful uses of a transparent fixture with a downlight was in an installation I did for a client who lived in a rural area on a lake. He traveled around the country and was home only on guest-filled weekends. Part of the weekend's entertainment included a hot tub built into the deck next to the lake. His electrician had installed several security floodlights, which automatically turned on each night illuminating the deck, whether it was the weekend or not. Whenever he was at home, his neighbors would paddle by, craning their necks to see who might be in the tub. The solution was an override switch for the security lights and two wall fixtures with downlights. The top fixtures were altered to accept candles instead of bulbs, giving very soft light for hot-tub time. The bottom fixtures were left as downlights, giving bright directional light below for other times. Now the neighbors cannot see unless they paddle right up. None have done so.

Preventing Light from Spilling Into the Neighbor's Yard

Make sure that outdoor light does not infringe upon neighbors by spilling into their yard or windows. In some areas, municipal codes prohibit such infringements. In all areas, positioning outdoor lighting to keep the illumination where it belongs. Light can be confined by burying floodlights in the ground, hiding them well in trees, or mounting floodlights at least 16 ft (4.9 m) above the ground and aiming almost straight down.

Bury floodlight in the ground.

Steps need light.

Walkway lighting.

Walkway lighting.

Pendants.

Post light.

Reassurance for Moving Around

Light can help people to change levels and direction safely. Light the areas for change—steps, walk and driveway junctions, and ramps—not the person. Client after client has proudly shown me his or her do-it-yourself outdoor lights aimed directly at the incoming guest. This lighting practice is an error. Navigation is hindered.

At steps, the lighting could be:

- Recessed step light, using a 25-watt standard A bulb. If the steps are masonry, owners must plan ahead to include the fixture; if the steps are wood, fixtures most likely can be added.
- Semirecessed or low walkway lights, adjacent to one or two steps.
- Overhead floodlight (10 to 20 ft, or 3 to 6 m above the steps), using a 65- to 120-watt PAR bulb.
- Post light adjacent to the steps.

Renters can install low walkway or overhead floodlights temporarily with waterproof cords and connections. Owners can install all kinds permanently.

At walkways, the lighting could be:

- Incandescent or deluxe-mercury fixtures on a

post or stake, about 2 ft (0.6 m) above the ground. The posts should be positioned so that the size of the dark areas and the lighted areas are equal. However, if continuous light is wanted, the spaces between the posts should be around twice the height of the posts, unless the manufacturer indicates otherwise.

- Overhead floodlighting.
- Chain-hung pendants, either in regular or low-voltage, hung from a tree or an eave, lighting about a 2-ft (0.6-m) area for each foot hung above the ground.
- Wall-mounted fixtures alongside walks or steps. The fixtures should be mounted at eye level (about 5 ft 6 in., or 1.7 m) unless they interfere with moving about.
- Ceiling fixtures in roof overhangs beside walks. The minimal area lighted by this type of fixture is usually about 4 ft 6 in. (1.4 m).

At junctions, such as walks and driveways, walks and steps, streets and driveways, the ing could be:

- Post lights mounted at eye level with well-shielded bulbs or low-wattage bulbs. Such posts can illuminate about 25 ft (7.6 m). The light from transparent fixtures should be equal

All drawings on this page are reprinted by permission of Kim Lighting.

to candlelight. Some post lights use candles for light sources; they are very distinctive for short-term illumination.

- Pole lights, such as small streetlights found in urban historic districts, lighting a greater distance. Very tall poles with security lights light even more. They can be obtained from most utility companies. They use either mercury or sodium bulbs. Think twice before using them. The color of the light is probably different from the other outdoor light. Choose them only to illuminate a large area inexpensively. They are bright. Do not install them on the bedroom side of a residence; the light is difficult to keep out. Consider the lowest wattage available; higher wattages are not necessary.

Reassurance in Parking Lots

For parking lots, the greatest amount of light over the largest area is produced by sodium lights on poles. The other choices, in order of efficiency, are metal-halide, mercury, and incandescent. Ordinarily, a parking lot that requires 400-watt mercury bulbs can use 250-watt sodium bulbs. They consume 37 percent less electricity and yield 40 percent more light than do mercury bulbs. Overall, sodium costs about half as much as mercury to operate, and mercury street lights are being replaced with these more efficient sodium bulbs. Sodium gives a golden glow similar to our image of sunshine.

Manufacturers supply technical information about the spacing of their poles, which depends upon the height of the poles and the characteristics of light distribution of the fixtures. For example, the distance from the outside edge of the parking lot to the first row of poles with one type of sodium fixture can be twice the fixture's height. The distance from an inside row of poles to another inside row should not be larger than four times the pole height. Therefore, if the poles with sodium fixtures were 20 ft (6 m) tall, the first row of poles could be up to 40 ft (12 m) from the outside edge of the parking lot. The distance from the first row to the next row of poles could be up to 80 ft (24 m).

Maintenance of parking lot fixtures is as important for long-term good lighting as is proper positioning. Clean parking lot fixtures every one to two years, and replace all bulbs every 4 years.

Automatic Reassurance

Outdoor lighting can turn itself on and off by a photosensor or a time clock. A photosensor turns the lights on automatically when the daylight gets dim—sometimes on dark rainy days. It turns them off again when the daylight gets bright—sometimes never during the winter, because the sky never gets bright enough. Unfortunately, the sensor cannot distinguish between cloudiness and sunset. Therefore, install a sensor where it will have the greatest exposure to the sky. Some posts or fixtures come with sensors; some do not. Sensors can be installed anywhere an electric wire can go.

I have done most of my outdoor lighting designs in the southeastern United States, which is heavily forested. Most of the lightposts have been overhung by trees and the sensors have been installed at the roof to gain enough exposure to the sky. Otherwise, the lights would turn on well before sunset and turn off long after sunrise.

For greater control, install a wall switch inside the building to override the signal from the sensor. The switch gives additional options. The sensor can turn on the lights after dark before anyone arrives home, and the lights can be turned off with the switch at bedtime, even though the sensor says it is still dark.

A time clock also activates and deactivates outdoor lights at predetermined times and days. However, compensations must be made for daylight savings changes by either resetting the clock or purchasing an automatically compensating clock. A time clock tends to be more wasteful of energy than a well-positioned photosensor with an override switch.

Parking lot lighting.

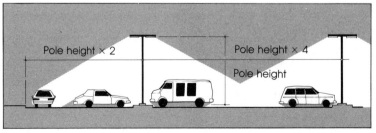

Decisions About Outdoor Lights

Decisions about outdoor lights depend on many factors, but the most important is placement. Almost any amount of light located in the right place is sufficient. Of course, if the surroundings are bright, like an urban or suburban setting, a lesser amount of light will be needed than if the surroundings are dark. Next, the size of the area to be lighted governs the choice of equipment. Large areas require equipment to spread the light; small areas require confining the light. Installation requirements also affect choices of lighting equipment. For example, if outdoor steps are to be lighted and are located 10 ft (3 m) away from a one-floor residence with no adjacent trees, the possibilities of equipment are limited to those that can be held up by a post. If the steps are adjacent to the house, wall lights could be mounted to illuminate them. Or if a three-story commercial structure on a remote truck route is to be lighted for security, fewer footcandles will be required and will need to be broadly spread over the front, back, and sides. Tall poles from a local utility company may be sufficient and aesthetic if well hidden among trees.

Always illuminate as large an area as possible with a low level of light. Do not illuminate small, scattered, bright areas next to dark areas. Illuminate large walls, driveways, and paved ground surfaces to help spread the light, creating a background against which intruders can be seen.

reassurances inside

Lights inside also provide reassurances. They brighten up the night to keep away whatever comes out in the dark. Both the very young and the very old feel more reassured with a light on at night. Also, evening or overnight guests can move around in an unfamiliar space undirected with the aid of low-wattage light. Low-wattage light does not have to come only from wall-plugged 4- or 7-watt bulbs (nightlights). Other fixtures can act as nightlights also and can provide additional functions at other times. Other fixtures are:

- A 30-watt reflector bulb downlight, in a bay, a bow, or a boxed window, over a buffet, over the front edge of an easy chair, and other places.

- Any fluorescent lighting system, such as cornice, cove, valance, or soffit. These systems light a larger area, but with energy efficiency.
- A 15-watt decorative ceiling or wall fixture in a bathroom. Such a fixture can enhance the interior style, such as an electrified doctor's lantern hung on the bathroom wall, and can identify the space for guests or customers.

- 75-watt reflector bulbs in down- or uplights to light the space between the window and sheer or unlined drapes to prevent anyone from seeing inside. This method is especially good for late-night office workers and those at home alone. Owners can use permanent downlights; renters can use portable uplights. In order to be effective, the brightness between the drapes and the window must be greater than the brightness in the space itself. The light obscures the view by creating an area of brightness in front of the drapes.

Light helps provide reassurance both inside and out. Outside, almost any amount of light located in the right place is sufficient. Inside, the amount can also be low and can be used as an additional lighting system for other reasons at other times.

light
to create
a view
16

Structures are viewed from the outside by many people. Unless the outside light is brighter, at night the structure is primarily seen by the light coming through the windows. Thus, in residences or commercial structures where the architecture contains special windows featuring unusual shapes, the exterior impression is enhanced by the light in these windows.

bay, bow, and box windows

If windows project handsomely from a structure, bay, bow, and box windows can attract attention. The light commonly used in the room containing such a window may be sufficient to illuminate it. But if the room is not used for the major portion of the night, the special window will be unlighted. It should be lighted, nonetheless. The amount can be very small, unless, of course, the outside is lighted like the blaze of noon. The amount of illumination needed in the window to capture any attention is three times the amount outside.

Owners who are building can prewire for a small, recessed incandescent downlight with a 25- or 30-watt reflector bulb. The light enhances the finish on the windowsill—stained wood, or covered by a cushion for sitting. At the same time, the light enhances the uniqueness of the window from the outside. Renters can temporarily install either a pin-up containing the same size incandescent bulb (25- or 30-watt reflector) or a throwaway fluorescent strip, whichever can be hidden best.

rose, arched, or other similar windows

Round, curved, or other specially shaped windows can be dynamic nighttime attractions that symbolize the structure. These windows can reflect light from a pale-colored surface, illuminating the window's form. The pale surface can be lined drapes, fabric stretched in a frame, or a painted plywood backing. The light source can be a fluorescent strip, cool in both color temperature (slightly bluish) and amount of heat, spreading illumination well over the reflective surface. Install two strips, one at the bottom and one at the top, or on both sides, wherever they are hidden best.

Otherwise, these windows can be lighted by reflected light from the room. If an arched window is above a door, the light from hallway

A lighted rose window symbolizes the structure.

fixtures, table lamps, or ceiling fixtures is enough. Likewise, the view of a shimmering chandelier could show through the arch and augment the beauty of both. The view of any visible fixture creates an impression of richness and high-quality ambience, desired by some and not by others. If you like to show and tell, do not miss this opportunity to discreetly flatter yourself.

clerestory windows

Clerestory windows are windows near the top of the wall, well above eye level. They can create an exterior impression and also a pattern of light on the outside. More than likely they can be illuminated by lights already in the space. However, sometimes a series of clerestory windows march across the face of a structure in various rooms. If so, make sure that all of the windows are equally and simultaneously lighted, for the best visual impact.

skylights

Skylights transmit light from the inside out as well as from the outside in. If the skylight is double-layered—a layer on the roof and a layer on the ceiling—fluorescent strips can be installed on the sides between the layers. Cover the bottom of the skylight with a translucent diffuser, letting light through but not being completely transparent. The quality of the light delivered is a moonlightlike glow.

If the skylight is a single layer only on the

A lighted clerestory window creates a pattern of light.

Bare bulbs create sparkle
and visual interest in a skylight.

roof, clear incandescent bulb fixtures can make it a lighting jewel, fashioning sparkle and visual interest for a sense of vitality in the space. The light will shine above and below. Because skylights are above eye level, fixtures and clear bulbs can be used. The fixture can be an inexpensive porcelain socket or a more expensive polished brass holder. The former costs around $2; the latter costs about $16. They should hold a clear G bulb of the largest size but should have the smallest wattage possible. The fixtures should be spaced evenly along two sides of the skylight; placing them along four sides is not usually necessary.

the view from inside to outside

When viewing the outside from the inside, you see either a dark void, if unlighted, or an extension of the inside space, if lighted. Light outside must be sufficiently intense to overcome the inside brightness. At the same time all inside lighting fixtures facing the window must be shielded, and the surfaces close to the windows that might reflect light (such as walls and floor) must minimize reflections by being low in brightness. Otherwise, the reflections on the glass will obscure the view to the outside. Balancing the relationship of interior and exterior brightness is important for lighting design—and is not often discussed. Trial and error is the best method, unless you hire the skills of an experienced lighting consultant. If you like trying and do not mind doing it over again, do it yourself.

There are many possibilities for outdoor lighting. (See Chapter 15.) In addition, residential owners and renters have many possibilities for lighting with the popular low-voltage outdoor lighting system package. It is designed to be installed by the purchaser with a screwdriver and hooked up with the convenience of a cord and plug. The package includes low-voltage waterproof cable, transformer to reduce the electrical current to a shock-free 12 volts, stake-mounted fixtures, and bulbs. These systems are supplied with 18-watt bulbs. However, within the limits of the transformer, a range of other sizes—7 to 36 watts in both spots and floods—could be used for greater or lesser punches and spreads of light. The transformer's capacity is limited. It determines the amount of usable watts and therefore the total number of fixtures. Sometimes the fixtures are packaged with col-ored lenses. Resist them. Order clear ones and keep the light natural looking.

Commercial owners and renters, on the other hand, need to have more substantial outdoor lighting that can take wear and tear. The best commercial outdoor fixtures are cheaper over the long run. At the least, commercial outdoor lenses should be shatterproof plastic, and the fixtures should be made of weather-resistant finishes and should be well secured to the structure.

daylight
for a
gloomy room

17

Do you have a room that is dark and gloomy? Unfortunately, in most residences—apartments, row houses, and single-family homes—one room seems to be forgotten by the daylight. During the daytime there is not enough light to perform simple functions. Electric lights get turned on and often left on, using electricity and sending utility bills higher than needed. Electric lights were invented for use after sunset, not after sunrise. Why not get the sunlight into a gloomy room?

Daylight comes from the sun as energy in the form of light. It comes either directly from the sun or indirectly from the sky; it is free. Using it as a source of free illumination is more important today than ever. Free illumination helps to conserve our precious nonrenewable resources.

See for Yourself: Do You Have Enough Daylight?

1. Can you see the sky clearly through the window from where you are, seated in a chair while reading, eating, or working?

2. If you cannot, you do not have enough daylight to see most tasks.
3. If you cannot, identify why.

Do not consider leasing or purchasing a home or apartment without applying this test. It can save you from living in a gloomy place. Often the causes of too little daylight can be corrected. If you know the causes, you can apply the cures.

CAUSE: ARE YOUR WEATHER CONDITIONS CLOUDY OR SMOGGY?

Clouds and smog can block the sunlight. Reflected light needs to be received from the sky to get as much daylight as possible. Therefore, if you live in a cloudy climate or under smoggy conditions, use the rule of thumb for daylight.

Rule of Thumb for Enough Daylight
Have an unobstructed view of the sky from where you perform any activity that requires seeing.

Clouds obscure the sun.

Can you see the sky clearly?

If not, you probably do not have enough daylight.

CURE: REARRANGE THE FURNITURE

Poorly arranged furniture can prevent you from getting sufficient daylight. Rearrangement can be the cure. Put desks, reading chairs, or any daytime work surfaces where they can take advantage of available light—but not glaring sunlight—from the window. Then daylight will be effective for illuminating whatever is needed to be done.

Desks and other tabletops can be placed parallel or perpendicular to windows.

- If you are right-handed, place the desk so the light comes from the left.
- If you are left-handed, the light should come from the right.

- If you are ambidextrous, take your choice.
- Never put your back to the window, because you cast a shadow on your work.

Place *reading chairs* so that the window light comes over the shoulder. Never face a window that could be very sunny, and never put your back to a window that is yielding very little light.

Beds can be placed at windows to gain daylight for those who like to read in bed during the day or for those who need to be in bed. For instance, does your teenager do homework on the bed with no concern for the amount of light he or she is getting? If so, put the bed at the window and have the sun supply light that nev-

When light is scarce, daytime work surfaces should be near windows.

A bed at the window permits
children to read with daylight.

Incorporate electric lighting
for nighttime.

er needs turning on or off. Integrate the bed to the window with a canopy headboard, enhancing the bedroom and at the same time making possible good, free light for reading. Combine draperies along the sides and a canopy board over the window as a headboard, and install a shade on the window for privacy.

Studying at night, on the other hand, requires electric light. Install an energy-efficient light with fluorescent tubes behind the canopy board for nighttime use. (Details for construction are in Chapter 5.)

In addition, you may want a pair of tall table lamps placed on either side of the bed and equipped with three-way incandescent bulbs (50, 150, 200 watts) or with fluorescent screw-based bulbs using less watts. Choose lamps with translucent shades to filter, not block, the light, thereby broadcasting it well. Or for children's rooms, double the amount of canopy lighting and omit the lamps. These arrangements are both energy-wise and foolproof.

CAUSE: ARE YOUR WINDOWS TOO SMALL OR NONEXISTENT?

For minimal daylight in any room, the window area should be equal to 10 percent of the floor area. Therefore, a 10- by 10-ft (3- by 3-m) room (100 sq ft, or 9 sq m) needs 10 sq ft (.9 sq m) of window. More area will give more light, and under cloudy or smoggy conditions, the sun's heat will probably not be too much.

CURE: ADD MORE WINDOWS

For the best daylight, consider increasing the number of windows, rather than just increasing the size of one window. Position windows on different walls. The more walls with windows, the more constant the daylight will be throughout the day. Likewise, glass in or adjacent to a door is considered a window and should be utilized whenever possible.

CURE: USE SKYLIGHTS (FOR OWNERS)

Owners can use a skylight if they have access to

Use glass in and around doors.

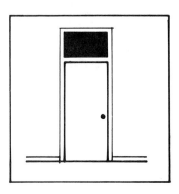

the roof. Skylights are more than just holes in the ceiling. They are windows to the changing panorama of the sky—fast-moving clouds, tree-tops, changing sunlight and, at night, a handful of stars and the moon. Skylights gather all the light possible. They are made of clear or non-clear glass or plastic. Some are equipped with interior screens, exterior awnings, built-in roller shades, or ventilation flaps. Some are domed; some are flat. Some are a single layer; some are double and sealed.

The general tips for skylights are:

- On sloped roofs, position skylights preferably on the north slope of the roof. Since skylights gather so much light, north light is softer—and better. Artists have known this fact for many years. If they are not possible on the north slope, then, in order of preference, east, south, and west are suitable. Skylights sloped to the south act as solar heaters, warming the air.
- On flat roofs with clear skylights, install a shade or other device to shut off the sun when it becomes too hot or too glaring.
- Choose a nonclear skylight for rooms that contain carpeting, wallpaper, and fabrics that might fade, particularly in blues and purples.

- Choose a clear skylight for rooms that do not have delicately colored finish materials or art-work subject to fading, such as kitchens, dining areas, atriums, hallways, bathrooms, and enclosed porches (Florida rooms), and any living area with mostly hard surfaces.
- Use nonclear skylights when the view should be obscured; for instance, on flat roofs where leaves will pile up.
- Use nonclear skylights to soften too much direct sunlight.
- Choose a nonclear skylight where the brightness impact on the ceiling needs to be minimized.
- A clear skylight can be finished by boxing in the space between the ceiling and the roof. The boxed-in part will be as deep as the distance between the finished ceiling and the roof.
- Those people that do not like a deep boxed-in effect should finish off the ceiling with a flat nonclear plastic or glass diffuser, not interrupting the surface of the ceiling.

Specific tips for clear skylights are:

- Even in cloudy or smoggy areas, some days are sunny. On these days, sunlight through clear skylights can cause blinding brightness and in-

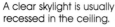
A clear skylight is usually recessed in the ceiling.

A diffuser on the ceiling eliminates the boxed-in look.

terfere with activities. Position the skylight so that the direct sunlight will be where you want it and when you are most likely to be in the room.

- In a kitchen, a skylight toward the west will admit glaring afternoon sun, just at the time the evening meal is being cooked. A skylight facing east or north would receive only reflected light in the afternoon. If a western orientation is the only choice, use nonclear glass or plastic to soften the light.
- In bathrooms, a skylight toward the east might thrust glare on the mirror in the morning. Make sure the sun's rays cannot reach the mirror. Strong, but not glaring light is usually welcomed in the bath.
- In bedrooms, a skylight toward the north is best. Toward the east will bring in early morning light that might disturb your sleep. South or west exposures can bring in strong sunlight that can easily fade your carpet or other materials and heat up the space too much in the summer.

In spite of the cautions, clear skylights add a visual dimension to a room not possible with a window. The passing clouds and the moon can be seen through them. On cloudless nights, moonlight is bright enough to allow you to move around without turning on any electric lights. In the winter, moonlit nights are brighter than early mornings.

CURE: INSTALL GLASS DOORS

Usually owners do not have access through to the roof for a skylight. However, if owners have a solid door in the room that faces a well-lighted space (interior or exterior), they can replace the solid door with a glass door. The glass door acts like a window, provided it is opposite brightly lighted windows in another room (windows facing south, preferably); it is in a bright, sky-lighted hallway; or it is an outside door.

There are more varieties of glass doors, including folding glass doors, than there are varieties of wood doors. Most glass doors have at least a wood or a metal frame. Do not forget the all-glass door, usually used in commercial buildings, but nonetheless suitable for residences. It is sophisticated-looking. An all-glass door becomes a full-length window. Privacy, if needed, can be maintained by choosing patterned rather than clear glass. It is a successful solution for rooms that do not have any windows.

Renters are usually unable to change the structure they lease, by virtue or vice of their property rights—the right to live in but not to alter without permission. Renters must rely for the most part on internal changes to augment daylight. (Review the sections on furniture arrangements and interior surfaces in this chapter.) However, a wooden door could, for the term of the lease, be replaced with the renter's wood and glass door, if it is hinged in the same way. Then, the renter could enjoy the daylight and reinstall the solid door before moving out.

CAUSE: IS THE SUNSHINE ON THE OTHER SIDE OF THE HOUSE?

CURE: CREATE EXTERIOR ARCHITECTURAL SURFACES

If you are the owner, create a large exterior surface such as a fence, wall, or a patio floor to redirect the sunlight into your room. Exterior surfaces, such as patios, pavement, the ground, fences, and walls, can reflect almost half the light they receive through the windows into the adjacent room. On sunny days the amount of direct sunlight striking these surfaces is more than that needed for interior use. On cloudy days the light reflected from the ground can be brighter than the sky itself.

Replace a solid door with a glass door to borrow light from another space.

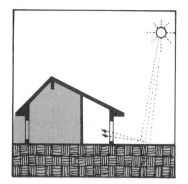

Light can be reflected.

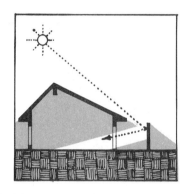

A fence can redirect the light.

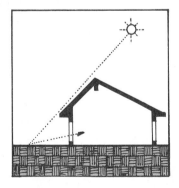

Create a surface to reflect light.

Is the sun on the other side of the house?

A concrete patio surface reflects light
through the windows.

144

Rules of Thumb for Exterior Reflecting Surfaces

The closer to the interior space, the brighter the light; the farther away, the softer.

The more white in the color of the surface, the brighter the light; the more black, the gloomier the light.

Be aware that pure or off-white (highly reflective colors) on these surfaces might produce glare. Pale to medium color values are reflective enough. Select the color among these values. Light will be redirected and usable. For example, a fence stained a rosy tan would reflect 30 percent of the light. A beige-colored brick patio would reflect at least 48 percent of the light back into the adjacent room. Both have pale to medium values.

GROUND REFLECTING MATERIALS

TYPE	AVERAGE REFLECTANCE
Natural Minerals	
pebbles	13
sand	30
water	95
bare ground	7
slate	8
Natural Vegetation	
dark green grass	6
other ground covers, ivy, etc.	25
Man-made Materials	
concrete	40
asphalt	7
brick—pale tones	48
brick—dark tones	30
wood deck	40
plastic grass	45
white terrazo	45

CURE: CREATE INTERIOR SURFACES

Colors on the interior surfaces can cause a room to be gloomy, and also they can cure it. Renters and owners can enhance the daylight by creating interior surfaces that redirect the light. Walls, floors, and ceilings become the most important control available. They need to be finished in colors that reflect as much light as possible. Pale colors with large amounts of white reflect as much as 85 percent of the light within the room. Dark colors, on the other hand, absorb so much that sometimes less than 10 percent is reflected.

For a gloomy room, the back wall (the one opposite and usually the farthest from the window) is particularly important. It adds or detracts from the total room illumination. A back

On gloomy days and at night, use an energy-efficient cornice light on the back wall.

wall receives little direct light, even on sunny days. The light it receives is already reflected from somewhere else. If the color on the back wall is highly reflective, the whole room can feel bright and airy. To increase the apparent light, use pale colors, mirrors, metallic wallpaper, or polished tile on the back wall. Create reflected light while making an interior design statement.

In addition, for very gloomy days or nighttime, equip the back wall from end to end with a fluorescent cornice light. The cornice will create a cheerful atmosphere by spreading the energy-efficient light where needed. It will heighten the colors and texture of the wall covering and enhance nearby furniture. (Review cornice light in Chapter 5.)

Never paint a ceiling in a gloomy room a dark color; do it only in a sunny room. Ceilings send back the light that is reflected from the ground, and gloomy rooms need all the light they can get.

The floor receives light from the sky, so keep the floor covering as pale in color as possible, reflecting back as much light as is available.

Other walls should conform to the average recommended for interior surfaces.

**Rule of Thumb for Average
Reflectances of Interior Surfaces**

The average reflectance of interior surfaces should be between 45 and 60 percent.

A wide choice of colors is contained within these minimums and maximums. This range does not limit the color treatment of surfaces, particularly walls, to white alone. Pale values of red and green, for instance, are comparable to the values of off-white in light-reflecting ability. Do not fixate on only white.

How do you know what the light reflecting value of a color is? Technical paint samples list the values as light reflectance percentage (LR%). The percentage is light thrown back; the percentage missing is light absorbed. Such color samples can be compared to the color of wallpaper, fabric, furniture, and other interior surfaces, indicating approximate light reflectance.

However, without the samples, color reflectance values of common items may give you some measure for judging colors. The color of a ripe tomato has a light reflectance value of 25 percent; the color of pine needles, 20 percent; the color of French vanilla ice cream, 80 percent; the color of butterscotch sauce, 60 percent; and the color of leaves on the trees in spring, 40 percent. If the color you are considering is about the same as your morning coffee grounds—13 percent—it is too dark.

MINIMUM AND MAXIMUM REFLECTANCE VALUES
FOR INTERIOR SURFACES
FOR GAINING DAYLIGHT IN INTERIORS

	MINIMUM	MAXIMUM
External		
ground	20	70
vertical surfaces	25	40
Internal		
ceilings	80	95
other walls	40	90
floor	20	70

CAUSE: ARE EXTERNAL ARCHITECTURAL
OBSTRUCTIONS BLOCKING THE DAYLIGHT?

Architectural obstructions that block the daylight are usually adjacent buildings, fences, walls, and large roof overhangs.

CURE: ALTER THEM IN SOME WAY

Owners who have external obstructions can alter them if they own them. If the obstruction is

Cut out a roof overhang.

removable—a roof overhang—cut it back or open it up over the window, admitting light. If the obstruction is not removable—a building, a fence, or a wall—paint it a value that reflects light—a color with a lot of white in it.

Renters who have external architectural obstructions cannot alter them structurally but might be able to repaint. Also, renters can rearrange their furniture and cover their interior surfaces with highly reflected colors to gain all the light possible.

CAUSE: ARE EXTERNAL LANDSCAPE MATERIALS
BLOCKING THE DAYLIGHT?

Landscape materials that obstruct are large trees, overgrown shrubs, and vines climbing on the structure.

Pruning a dense tree . . .

. . . permits light to enter.

CURE: ALTER, REMOVE, OR REPLACE THEM

Alter landscape materials judiciously. Reshape but do not disfigure them by pruning. Remove or replace those that cannot be pruned satisfactorily. Replace evergreen materials with deciduous materials that drop their leaves in the winter and grow leaves in the summer. Replace broad, dense trees and shrubs with open, smaller-leafed ones. They give airy shade and some light. A Japanese maple, for instance, gives soft shade, drops its leaves in the winter, and is an attractive shade tree, not too dense, as compared to a sugar maple, which is broad-headed, large-leafed, and very dense.

Owners who cannot alter landscape materials need to consider using skylights or glass doors and altering exterior architectural surfaces as suggested in this chapter to enhance the daylight. Renters who cannot alter the landscape materials need to use furniture arrangements and interior surface changes included in this chapter. However, if the landlord pays the electric bills, it might be possible to motivate him or her to assist in getting good daylight with other cures, so that the cost of utilities could be reduced.

keep the
hot sun out
but let
the light in
18

Even though sunshine makes us feel good, too much sunlight in our interior spaces can have adverse effects. It can bring in too much heat and be too glaring when it is direct. Sometimes it restricts the use of furniture and activities, such as preventing people from sitting in a particular chair or on one side of a table. Likewise, sunlight can fade upholstery fabric, blister furniture, and bleach carpets.

Can the light be kept and the heat and harshness be shut out? Yes, the sun's light can be obscured or eluded. Obscure it by blocking or scattering the light by means of window treatments, architectural devices, or landscape materials. Obscuring methods can be applied both indoors and out. Elude it by getting away from the direct light by means of furniture arrangements or dark interior finishes. Eluding methods must be applied indoors.

choosing a sun-control method

What is your status—owner or renter? If the space or the structure is not owned, it cannot be altered, at least not without permission.

What is the height of the structure? Outdoor obscuring devices—trees, fences, and others—can be used with a one- to three-floor structure (low-rise). They cannot be used with a four or more floor (high-rise) structure, unless the device originates from a balcony (potted tree).

What is the location of the sun? If you know where the sun is at various times of the day, you will know where to expect direct sunlight. In the early morning the sun is low in the sky, rising to the highest elevation at noon standard time, descending again and setting at the end of the day.

All the time the sun is rising and descending, it moves from an easterly direction in the morning to the south at noon, to a westerly direction late in the day. The arc of that movement is larger in the summer than in the winter. In the summer, the sun actually rises north-northeast and sets north-northwest; in the winter it rises east-southeast and sets west-southwest. Consider this information in relationship to the windows in the room.

What is the angle of the sun? Knowing the angle above the horizon at various times of the day permits you to determine the position for a sun-control device. The sun is at low to medium angles in the east and west. The sun is the high-

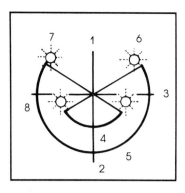

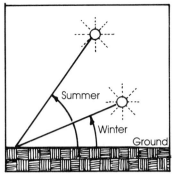

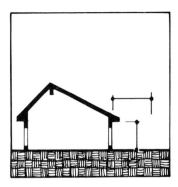

Sun's arc. (1) N, (2) C, (3) E, (4) winter arc (5) summer arc, (6) rise, (7) set, (8) W

Angle of the sun.

How high and how wide should a sun control be?

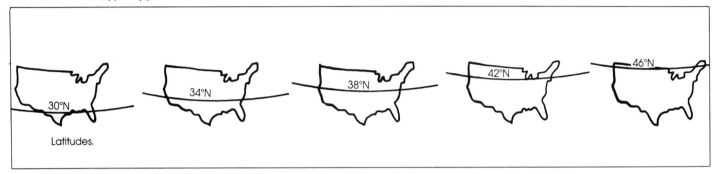

Latitudes.

est in the south at noon each day, but the angle is different for each latitude. To determine the angle of the sun, know your latitude. In lower latitudes the sun is higher in the sky; in higher latitudes the sun is lower.

At the same time, the sun is the highest in the sky in the summer, whatever the latitude, and lowest in the winter. All in all, it is easy to remember that the sun is highest at noon in the summer at south windows, and at the east and west windows it is essentially low in all seasons.

Do not try to build, buy, or rent a sun-control device before you know three pieces of information: how high a sun-control device should be, how far it needs to stick out, and what the most effective position is. Determine this information by pretesting it with a pencil and paper.

Pretest Sun-control Devices Yourself

1. Determine the nearest latitude. If you are halfway between two latitudes, divide the difference in half, adding the half to the lower number. Thus, if you live in Philadelphia which is halfway between latitudes 38° and 42°, the summer angle would be 67½° and the winter angle would be 24°.

2. Make a small-scale drawing of the room and the surrounding walls, like a drawing of a doll's house cut in half, and a drawing of the device being tested—a fence, a tree, an overhang.

3. Using a protractor, put the center point at the edge of the sun-control device—the edge of the roof overhang, the edge of an awning, or the lower and upper edge of a tree.

4. Project the angle of the sun at your latitude for both summer and winter, directly into the room. Everything on the far side of the line from the sun would be in shade; everything on the near side of the line toward the sun would be in the sunlight.

Use a protractor . . .

. . . to test a sun-control device.

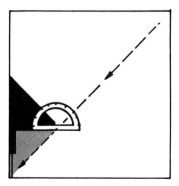

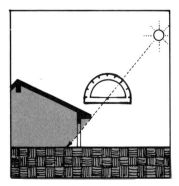

This pretest will show the summer's sun penetration (the least) and the winter's sun penetration (the most). The effectiveness of a control device can be determined before investing in it. Is a trellis over a window going to give enough additional shade? Is a medium-sized tree tall enough? Who wants to build a deck roof only to find that it is too short to give shade? Likewise, this kind of information permits pretesting eluding methods. It is truly a handy tool to ensure that money is spent wisely and effort is spent efficiently.

What is the type of activity in the room? Sedentary activities are more likely to be disturbed by harsh sunlight than nonsedentary ones. For instance, sitting at a desk and reading social security forms in harsh, hot sunlight would be more uncomfortable than taking inventory of the supply cabinet. Likewise, many short-term activities, such as brushing teeth, are rarely affected by too much sun. Therefore, sun control is not as necessary. On the other hand, many long-term activities are affected by too much sun and do require sun control.

window locations

Windows are the only way that sunlight can enter a structure. Their location is important, and the direction they face determines when they receive sunlight.

Windows facing the east receive direct light in the morning at a very low angle. Devices are necessary for bedrooms for those whose sleep would be disturbed by light or for offices where people get to work early.

Likewise, west windows receive the sun directly in the afternoon in a decreasing angle till sunset. At that time of day it seems harsher and brighter. Therefore, sun-control devices are necessary for dining rooms so that dinner and glaring light do not come together, and for offices where late afternoon work might be impeded.

South windows receive sunlight between 10:00 A.M. and 2:00 P.M. In the summer, sun control is obtainable by outside devices that block most, but not all, the sun and reflect some light. In the winter, sun control is sometimes needed, but since the sun warms it is often welcome. Nonetheless, it is lower in the sky and penetrates deeper into the space. When heat is not wanted, daytime activities can be lighted by the sun, as long as sun controls are available.

North windows never receive direct sunlight. The light coming in a north window is reflected from somewhere else—the sky, the ground, or another building. North windows do not need any obscuring device, not even a roof overhang. More often than not, however, structures are designed with the same overhang all around. This practice ignores the sun's arc through the sky and indicates a need to equalize the roof—probably unnecessary.

How many windows are in a space and where they are located are important. If only one wall has a window, the control device for that window governs the sunlight for the whole space. If the device is heavy draperies and they are closed, the direct sunlight gets shut out along with all the reflected skylight. Frequently, after the drapes are closed, electric lights are turned on to allow everyone to see, even during the daytime. Therefore, the sun is controlled, but money is spent on electric light. If daylight is available, use it. In general, small rooms can manage with only one window for light, but larger rooms need additional walls with windows to get sun at different times of the day.

More control is possible if two walls have windows. When one window receives direct light, it can be blocked and the other window can provide reflected light. More often than not, people seated in such a room are likely to face one of the windows, and therefore some form of sun control is necessary. Moreover, sometime during the day, at least one of the windows will receive direct sunlight.

Similarly, rooms with windows on three or four walls require sun-control devices. At least two or three windows will receive sunlight sometime during the day. Therefore, any and all control devices are appropriate. For greater control, use several devices at the same time. (I have had only one lighting client that had a room with windows on all four walls; the space was spectacular, but the client was unhappy because the sun faded her wall-hung art. We needed to obscure the sun and not put in any electric lights that would fade the art further.)

Clerestory windows (windows on the wall above 8 ft, or 2.4 m) give good light, but the direct sunlight can penetrate and fill the room with glare. Indoor controls are difficult to reach but not impossible if well engineered to be managed from the floor. Light can also be controlled with outdoor devices and ceiling color.

Skylights gather the greatest amount of sunlight possible, spreading it throughout the room. Like clerestory windows, indoor controls are difficult to reach unless they are well-designed. Therefore, consider using a skylight with a built-in louver or purchase a motorized shade to control the sun. If neither of these controls is available, drape a cloth or use a large umbrella, some wooden baffles, or some landscape materials, inside or out, to soften the light. Never install a skylight without sun control in a room with surfaces that are prone to fading (draperies, carpet, and wallpaper). Instead, install in a room that does not require light control, such as a bathroom, hallway, kitchen, or any space that has mostly hard surfaces—tile floors, wicker furniture, and painted walls.

methods of control

Obscuring

Control the sun by obscuring it—blocking or scattering the light. Obscuring methods are window treatments, architectural devices, and landscape materials.

WINDOW TREATMENTS

A window treatment is putting something at a window to reduce the sun's glare and harshness. However, it will not reduce the heat unless it is insulated. A highly adjustable window treatment permits the light from any angle to be redirected, scattering and softening it. They are best for sun control. Each type of window treatment has different degrees of adjustability and control the sun at different angles.

Draperies are the most common window treatments. But they are poor controls if windows are only on one wall. Depending upon the thickness of the material, draperies can either block or scatter the light. Dense material blocks; thin material scatters. When they are fully closed, dense draperies block the sun low in the sky. When they are partially closed, they sometimes block the sun at oblique angles. Thin drapes cannot block the direct sun at all. If thin drapes cover the window at all times, make sure that electric light is not also being used during the day. Daylight is free, and it should be controlled in some other way.

Grilles, screens, and latticework scatter the sunlight coming from an angle, but in most cases they cannot be adjusted to control the low sun. This treatment works best when the sun is high.

Shades can block all or part of a window. When part of the window is blocked, the rest of it transmits light from the sky or ground. To accomplish control, shades can be pulled from the top or the bottom. If pulled from the top, they block the sun high in the sky; if pulled from the bottom, they block the low sun. Therefore, aesthetics or convenience alone should not determine which way to pull shades. Sun control should help.

Shutters may be louvered or solid. Louvered shutters can adjust like venetian blinds and scatter the light. Solid shutters block the light completely or partially, depending upon their position.

Venetian blinds are the most adaptable of interior window treatments. They can deflect sunlight and redirect it to illuminate the room. Light can be scattered up to the ceiling, down to the floor, or to the left and right, depending on whether the blinds are horizontal or vertical and how they are adjusted. Whatever the adjustment, blinds are the most versatile sun-control device. (Window manufacturers in Scandinavia make windows with the venetian blinds built in between the layers, which makes them dust-free.) Again, aesthetics alone should not dictate whether the blinds should be vertical or horizontal. The decision should include control of light.

The color of a window treatment affects the color of the light coming in through the window. Light picks up the tint of the object it strikes. Of course, it strikes the window treatment first.

See for Yourself: Does Light Pick Up Color?
1. Cut the bottom out of a shoe box.
2. Tape waxed paper over the bottom (not clear plas-

tic food-storage wrap but old-fashioned, semi-transparent waxed paper).

3. Shine a flashlight inside the box, aimed at one side.
4. In front of a mirror, observe the color of the light seen through the waxed paper.
5. Lay two pieces of bright-colored paper (red, orange, or deep purple construction, Christmas, or other kind of paper) inside the box along the sides.
6. Again shine the flashlight on one side and observe the color of the light. It will be tinted with the color of the paper you have put inside the box.

The best color for a window treatment is neutral or near neutral; the worst is an intense color that distracts or becomes a source of glare itself. Choose a color that gives no unwanted effect of its own to the interior space.

The construction of a window is very important because it creates sun control. Ideally, sun control should be accomplished by structural gradations from the brightness of the outside to the dimness of the inside. Windows in older homes created such a gradation. They were embedded in thick walls. They had deep sills, tapered wood between the glass panes, and movable shutters both inside and out. On the other hand, in contemporary buildings, windows are placed in thin walls, offering little to soften the light, especially in curtain-wall office buildings. Special attention must be paid to deliberately softening the light at and around windows. Softening can be achieved in contemporary construction by using window treatments and landscape devices as well as adding other architectural devices.

ARCHITECTURAL DEVICES

Architectural devices block and scatter the sunlight. They can be adjacent to, attached to, or part of the structure. They can be stationary, movable, or even removable. They give complete or partial shade, reducing the harshness of the sun for the inside spaces. In addition, heat is prevented from ever reaching the structure by devices that block the sun's rays. In hot weather, an air conditioner or fan does not need to remove the additional heat, and money is saved. External architectural devices can obscure the sun in different positions in the sky, but for the most part they are not adjustable.

Awnings can be either adjustable or nonadjustable. They can have side panels, blocking the sun from high, medium, and low oblique angles. Awning material has shorter life expectancy than the structure to which it is attached.

But they are decorative and add to the ambience, besides being functional.

Balconies block the sun for the window below, thereby becoming a roof overhang. Apartment dwellers in structures with balconies appreciate their upstairs neighbor's balcony more days of the year than the neighbor.

Fences, walls, courtyards, and other buildings block the sun, particularly the rising and setting sun. Moreover, they can provide privacy and reduce noise. The architecture of fences, walls, and courtyards can be visually, auditorily, and socially pleasing, while it provides sun control.

Sun protection and privacy from a fence.

Patterned glass and sunshielding glass or plastic cut down the sunlight in different amounts. Patterned glass diffuses the light unless the sun is aiming directly at it. Then, the light appears brighter than through clear glass, because the pattern augments and makes the light brilliant in the way that cut glass does. (Patterned glass is most often used in shower stalls.) On the other hand, sunshielding glass or plastic is either reflective or tinted. Reflective glass or plastic bounces back between 8 and 80 percent of the sun's light, depending upon its manufactured characteristics. Tinted glass or plastic bounces back 6 to 8 percent. However, direct sunlight is a problem for these sunshields, too. Question your supplier carefully to determine the exact amount and exact type of sun control you are purchasing. In high-rises, where other architectural options for suncontrol are limited, sunshielding glass or plastic is a must; in low-rises it can be very useful.

Latticework is a decorative form of crisscrossed wood or metal strips, which can be used to cover a window. Screens and grilles are usually metal. All three can diffuse the light somewhat, but none can block the direct sun, because of the openness of their construction. Scattering the light, in some cases, might be enough.

Louvers can be added to the outside of many structures to block the sun at the angle needed. They are suitable for most building styles, except traditional period styles like Neoclassical, Southern Colonial, and Cape Cod. Louvers have been used successfully for many years on residences in semitropical climates and on contemporary commercial structures in all climates.

Porch or deck roofs block the sun not only for the porch or deck but also for the adjacent interior space—two benefits from one device.

Recessed windows create a gradual gradation of light from the sky to the room.

Bay and bow windows function as recessed windows if the depth of the bay or bow is great enough to shade the interior space from direct sunlight.

Roof overhangs are permanent devices to block the sun as well as to keep the rain off the windows. On the east and west sides, overhangs cannot block the sun because it is usually too low in the sky. On the north, an overhang is not needed for sun control. (The sun never shines directly in north windows in this hemisphere.) On the south, overhangs are beneficial.

Shutters can be used on the sides of or over windows. Either way, they must be movable to be effective. Ideally, side window shutters should close completely and be controlled from the inside. Overhead window shutters can close completely but unfortunately can never be completely opened. Consequently, they always block the light from the sky but allow reflected ground light. Without a doubt, all fixed shutters are useless for sun control and are strictly decorative.

Trellises are cross-barred metal or woodwork. Trellises can be placed at right angles alongside the window or over the top. Trellises along the sides diffuse the light from an oblique angle (the rising or setting sun). Above the window, trellises diffuse the sun when it is high in the sky. A trellis can support vines or other landscape materials, adding a rich, softening effect besides giving the bonuses of color, smell, and beauty.

Which architectural devices are usable are determined by the type of structure and whether that structure is owned or rented. For example:

- Roof overhangs are not applicable for high-rises, unless you are the owner or builder and are able to specify overhangs on the building.
- Trellises, awnings, lattice, removable grilles, and louvers are excellent devices for renters of low-rises. They can be built, installed, and removed when moving time comes around.
- On the other hand, a courtyard, fence, or other building could be built by owners in low-rise structures to obscure western setting sun.
- Movable shutters are usable for owners, especially those who plan to build.

LANDSCAPE MATERIALS

Landscape materials—trees, shrubs, vines, and hanging plants—can block or scatter direct sunlight. These materials can be movable (a small tree in a tub on wheels), or they can be permanent for the life of the material (a planted

Tall trees obscure . . .   . . . the sun at mid-day.

hedge). Some materials always produce shade (evergreen); others produce shade only in the growing season (deciduous). Determine the height of landscape materials and choose those that can obscure the sun now and later. Small trees at eye level (around 5 ft 6 in. or 1.7 m) obscure the sun very low in the sky when rising and setting. Medium-sized trees (around 20 ft or 6 m tall) obscure it before midday in the summertime and during midday in the winter when the sun is lower in the sky. Tall trees obscure the sun high in the sky, at midday in the summer.

The broader the landscape material, the wider the area it shades. The denser the growth, the more sun it blocks. For instance, a small-headed tree shades only a very small area. A broad tree shades a larger area. In addition to providing shade, landscape materials can also provide privacy, enhancing your indoor and outdoor living.

At different times of the day, different types of landscape materials are best for obscuring the sun. Early morning and late afternoon sun protection can be gained by hedges and shrubs that grow compactly from the ground and are high enough to obscure the sun. Middle of the day protection can be gained by medium to tall trees with broad, compact growth, by

Broad landscape materials . . .   . . . block the sun.

Work surfaces, tabletops, and desks should not be next to windows that receive direct sunlight at the time of day when the surfaces would be used.

Sofas, chairs, and lounges should not face bright, glaring windows, particularly if the seat is bathed in the sun.

In a low-rise structure, furniture arrangements for eating should not require a seat to face an unprotected east window (at breakfast time) or an unprotected west window (at dinnertime). In a high-rise, the sun would probably be below the level of the west window at dinnertime.

vines or other plants growing on a lattice and hanging baskets. Tall trees also can protect high clerestory windows from the noontime sun.

Get a bonus by choosing landscape materials that produce flowers, particularly fragrant ones. For the fullest enjoyment, choose those landscape materials that bloom when the climate allows you to be outside or to have your windows open.

Unlike other devices, landscape material is living and has specific requirements for growth—light, soil, and moisture. Choose the materials whose requirements you can satisfy and whose susceptibility to disease is low. You will want to ensure that the device will be reliable over the years and will not need to be replaced.

Eluding

The sun can be controlled by getting away from its light. One method of eluding is to arrange the furniture away from the sunlight. The second method is to cover the major surfaces of the room with colors that either absorb some of the sunlight or blend with the bright light.

FURNITURE ARRANGEMENTS

Furniture arrangements are often the last method employed for eluding sunlight, but they should be used more often. In rooms facing south windows, use two different furniture arrangements—one for summer and one for winter. In the summer, the furniture can be closer to the windows but not close enough to be in the sun. In the winter, the furniture should be farther away, since the sun penetrates more deeply. With any arrangement or any window exposure, several rules of thumb should be followed to elude the sunlight.

INTERIOR FINISHES

All major interior surfaces—floor, walls, and ceiling—are finished in a color. Sunlight striking these surfaces will be absorbed and reflected back in the quantity allowable by the color of that surface. Dark colors (dark blue, black, deep brown) reflect only a small percentage of the light. Pale colors (beige, white, pale yellow) reflect most of the light received.

The floor receives most of the direct sunlight in any room. To reduce the amount of light, the floor can be made dark with wood stain, carpet, or other floor covering.

The wall containing the window is the most critical wall for eye comfort. If the window receives direct, unrelenting sunshine, this wall should not be dark. It should be pale so that no harsh contrasts are created.

Contrary to the usual practice, pure white walls should not be used where windows receive unrelenting sunlight. In sunny climates and in high-rises, windows are more likely to receive direct sunlight for long periods. Consequently, pure white on the wall is too bright. White creates an unpleasant brightness inside that does not subdue the outside brightness. Instead, use a healthy shade of pastel or a color that reflects only 70 percent of the light. There are many shades and colors to choose from in this reflectance range. How do you know how much a color reflects? Ask your local paint store for technical paint samples. These samples list light reflectance values, indicated by LR%. The percentage refers to the amount of light reflected; the amount remaining is the light absorbed.

Walls adjacent to the window receive sunlight. If the walls are dark, they become harsh contrasts. Finish them with a high light-reflecting color, but not necessarily white.

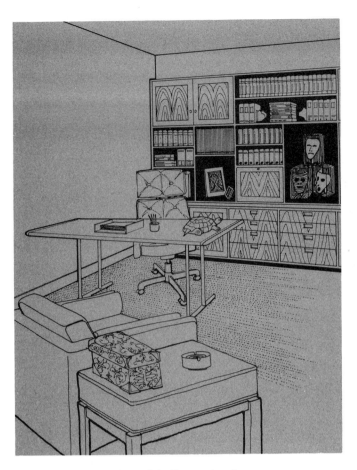

In sunny climates, put daytime work surfaces away from windows.

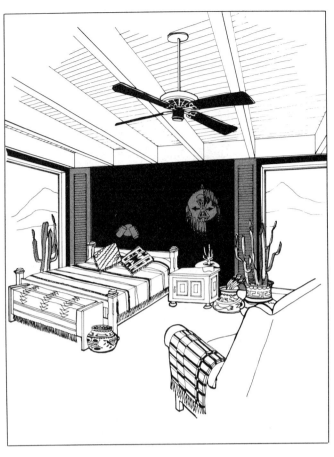

A dark window wall creates too much contrast in sunny climates.

The back wall of a room can be finished in a dark color. It does not contrast with the bright sunshine and helps to absorb excess light.

Ceilings also can be dark to absorb light, since direct sunlight reaches the ceiling only very early in the morning or very late in the day. Dark ceilings help control the light from clerestory windows if no other control device is possible.

FURNITURE COLORS

Any furniture, particularly a desktop or other work surface that receives direct sunlight, should be finished in a nonglossy, pale color to avoid eye-fatiguing contrast. But furniture away from the direct sunlight can be dark, thereby absorbing some light. (Mediterranean countries use dark furniture; for example, Greek cottage and country Italian furniture.)

using multiple-control devices

Sun-control devices should be used together. They can effectively moderate the light throughout the changes of the day, the months, and the year. They can work either simultaneously or sequentially.

Multiple devices work simultaneously to soften the light. For example:

Outside
- tree above the roof
- trellis adjacent to the window
- shutters and hanging plants inside the windows

Inside:
- window wall finished in a pale color

Multiple devices.

Simultaneous devices sometimes block and sometimes filter. The usable light is kept. Heat and harshness get shut out.

Multiple devices work sequentially, controlling the sun in its daily arc through the sky. For example, a family room, with windows facing east, south, and west and used all day long by a member of the family, could benefit from several control devices.

Outside Devices
- fence along the northeast and southeast lot lines
- group of small dense trees near the windows at the east to southeast
- awning on the south facing sliding glass door
- group of medium-tall dense trees and dense shrubs located away from the windows to the southwest
- carport on the west side

Inside Devices
- vertical venetian blinds at all windows
- ceiling painted a dark color
- furniture arranged to receive the best reflected light

Together these sun-control devices would regulate the sun for a single-family, owned home in a very sunny climate. Other types of structures, ownership, and climates would require a different set of devices.

the
best possible
desklight
19

Do you work at a desk for many hours? If so, 85 percent of the information received is transmitted through the eyes. Therefore, the lighting at a desk should maximize the ability to see information. Lighting is important; whether the desk is at home or at the office, it should be the best possible.

The best possible lighting is the best amount and in the best position to enhance vision. The amount of light required is influenced by many things, including readability and length of time required for reading.

Are you scrutinizing poor photocopies, with the paper and the writing about the same shade of gray? Are you trying to distinguish numbers that are small but important? Are you working with glossy technical journals or catalogs? Are you reading penciled longhand notes for hours on end? If you are performing these or other difficult visual tasks, more light is needed than if the tasks were sharp, simple, and unimportant. Sometimes, the amount that enhances vision is as high as 200 footcandles (usually the amount of light supplied for drafting and produced by four fluorescent tubes in an overhead fixture). It certainly is no less than 30 footcandles (usually the amount of light in a reasonably lighted office).

light as a hindrance

Light can actually hinder your work. It can be too little, be too much, or create a glare. If there is too little light, the eyes strain and become fatigued. In large offices, glare can come from the rows of ceiling fixtures themselves. The eyes must constantly adapt to the bright light above and the dimmer light below, as you look up and down and up and down. Likewise, if the work is too bright and the surroundings too dull, the eyes exert to adjust to both conditions over and over. If the light fixture provides light at less than a 30° angle, glare reflects from any glossy surface—photographs, glasstopped desk, highly polished wood, metal objects, or glossy magazines. Reflected glare produces a blur, destroying the readability of the printed material and causing you to move your head or the material. The blur is called a veiling reflection. In the long run, discomfort and loss of concentration can occur. In the office, these effects can reduce productivity and can cost money.

**See for Yourself: Do You Have
Reflected Glare at Your Desk?**

1. Put a small pocket mirror on your desktop, exactly where you read or write.
2. Can you see a light fixture in the mirror?
3. If so, you will receive reflected glare.

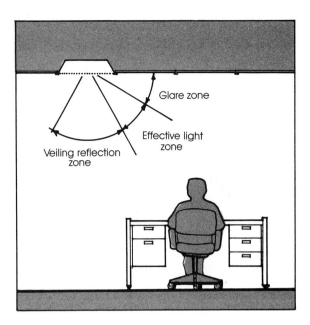

Reprinted by permission of Lightolier.

If there is any kind of glare, the eyes often do not complain and the physical signals are subtle. Glare adds to fatigue. It causes discomfort, usually slight, but sometimes considerable. Visibility is lost. Two measures have been developed and are being used to ensure the best possible lighting. The first is visual comfort probability (VCP), a method of predicting the comfort of a particular lighting system. The second is equivalent sphere illumination (ESI) a measurement of the amount of light in terms of visibility and used to compare lighting systems. However, these measurements may not have been available when your office was built or relighted. How can you gain the best possible light without them? Understand how glare, both direct and reflected, can be controlled, and control it by the best options at your disposal.

Rule of Thumb for Glare-Free Light

Glare-free light is received if the light comes from a very narrow angle, between 30 and 60 degrees from vertical.

Position the desk to receive light from glare-free positions:

- Never have the brightest light source—window, ceiling fixture, or lamp—above and forward of the desktop.
- Never have a highly concentrated, single light source directly aimed at the desktop.
- Do have a single, ceiling-mounted fluorescent fixture located above the head.

combining daylight and electric lighting

If the light sources are a window and electric lights, balance the daylight, which changes position throughout the day, with electric light.

- The window should be on the left for right-handed people and on the right for left-handed people—opposite the hand used for writing.

- If usable light from the window fades, replace it with electric light from the same side of the room.
- If light at the window is too bright, equalize it with brighter room illumination.
- If light at the window is too dull, illuminate the desktop with enough electric light.

at the office

Since the ceiling fixtures in an office are not movable, get the best possible light by moving the desk or the desk lamp. Light from fluorescent fixtures should come from either or both sides or from above the head to avoid glare.

**Office Ceiling
Fluorescent Fixtures**

If the desk is movable, choose one of these solutions:

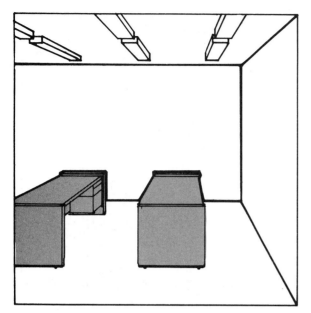

Rearrange your desk . . .

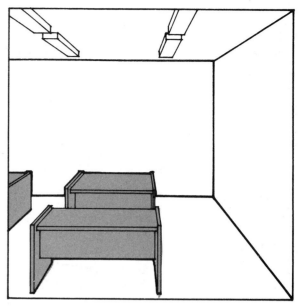

. . . between two rows of ceiling fixtures.

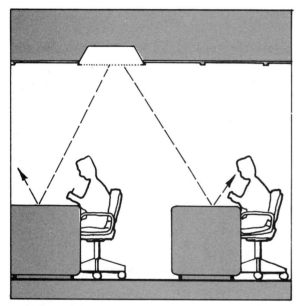

Light should reflect away from, not in your eyes.

- If many rows of fluorescent lights form a line on the ceiling, rearrange the desk so that it is between two rows of fixtures and the short ends of the desk are perpendicular to the long part of the fixture. The light will come from both the right and the left, balanced and shadow-free.
- If the desk is at the back of such a room, turn the desk around so that it does not face rows and rows of fixtures.
- If fluorescent lights are not in long rows, move the desk, so that it is parallel with the fixture and the fixture is directly over the head. The light will come and the glare will go away.
- Never allow a fixture to be directly in front of and above a desktop.

If the desk is not movable, change the working tools by using low-gloss paper and ink or ball-point pen instead of glossy paper and pencils, or convince the boss that productivity would be increased with glare-free light. Call his or her attention to the fact that the average worker's earnings, including fringe benefits, cost the company more than 200 times the electricity used by that company. Therefore, the best possible lighting can enhance productivity and save operating dollars. Further, because of its impact on office space, lighting could improve the most but cost the least. If you have trouble convincing the employer, get help from lighting consultants and light bulb manufacturers. They have the knowledge to solve lighting problems and have computer programs to analyze a lighting system's effects, costs, and benefits.

Furthermore, if work is done on a video display terminal, reflected glare is even more crucial because the area for glare is as big as the video screen and is continual. A video terminal should be moved so that no lighting fixture can be seen reflected on the screen. Also, when you look up from the screen you should not see any bright light sources. Your eyes need a rest. Judicious rearrangement of furniture or relighting with different light fixtures should be able to provide these conditions.

Fluorescent Lights on the Office Desk

If the desk lamp is fluorescent, make it advantageous:

With one fluorescent desk lamp, put it perpendicular to the person seated.

- With only one fluorescent desk lamp, put it on the side opposite the writing hand and position it perpendicular, even though it seems strange.
- Get a fixture with a special lens that redistributes the light in two opposite directions. It is called a bat-wing or bilateral ray lens. It avoids reflected glare, positioned parallel to your desktop.

- Put a pair of fluorescent lamps perpendicular and on both sides, balancing the light.

If the desk lamp is incandescent, see the lamps section in this chapter.

at home

A table at home sometimes becomes a short-term desk. A permanent desk also functions for work done at home—sometimes for business accounts and sometimes for household accounts. Wherever the work surface is located, it should be lighted sufficiently and comfortably by lamps or other lighting fixtures. At home a single light source is usually used to illuminate a desk surface. If so, its position and the amount of light it produces are critical. Likewise, the area around the desk is critical and needs no less than one-fifth as much light as the desk. Further away, one tenth as much is needed. As a result, the light at the desk would not be surrounded by darkness and the eyes would not have to work hard. To get one-tenth as much, turn on another light source near the desk. At the least, it should be one 4-ft (1.2-m) fluorescent tube or one 150-watt incandescent bulb within 5 ft (1.5 m) of the working surface.

Lamps

Lamps—floor, table, or wall—used as desk-lights can be either adjustable or nonadjustable, but they must deliver light well over the desktop. Therefore, desk lamps have five requirements:

- A shade that transmits light.
- The bottom of the shade 16 in. (41 cm) wide.
- The bottom of the shade level with the eyes.
- A diffuser, either a glass bowl or perforated disc under the shade. A diffuser improves the lighting quality by distributing it upward and outward. All desk incandescent lamps should have a diffuser, but at the least must have a diffusing incandescent bulb—coated, never a clear bulb.
- The ability to provide equivalent to 200 incandescent watts or 60 fluorescent watts for the desk. (For incandescents, use a 200-watt bulb, three 75s, or two 100s; for fluorescents, use two 30-watt tubes, a 32-watt, plus a 22-watt circular together, or two super energy-efficient 18-watt screw-based bulbs.)

Position a single lamp to avoid shadow—on the left side if right-handed, on the right side if left-handed. Further, the lamp should be adjustable if the work surface is adjustable, like a drop-lid desk, to accommodate to both the open- and closed-desk positions.

Ceiling Fixtures

Ceiling-hung pendants, chandeliers with downlights, and ceiling surface-mounted fixtures can also illuminate tables or desks. A single pendant, like a single lamp, needs to hang on the opposite side of the hand used for writing. A pair of pendants could be on both sides. A linear fixture (a library light) can hang centered on the tabletop, illuminating from both sides. As a result, two different people using the same desk can be satisfied. The same satisfaction is obtainable with a pendant adapted to a ceiling track and moved from left to right as required.

Chandeliers with downlights are normally hung in the center of a table and are not intended for extended use. But, for any work at all, choose a chandelier with a downlight that has the capacity to utilize at least a 50-watt reflector bulb. (See details on downlights in chandeliers in Chapter 3.)

In addition, a surface-mounted ceiling fixture with a lens is capable of spreading light. But position it above the head and never have your back to the light; it creates shadows. The most energy-efficient source is fluorescent, of

A pendant adapted to a ceiling track can move from left to right as required.

course. Avoid all open downlights; the light they provide is harsh. If they have reflector or PAR bulbs, radiant heat spills out. Therefore, if an open ceiling fixture is over the work table, use at least a standard A bulb, or better yet purchase a lens to soften the light. Lenses or louvers from the same manufacturer are often interchangeable with the same size fixture. (See lenses in Chapter 6.)

Sample Electric Cost

Compared to a 200-watt incandescent bulb, an 80-watt fluorescent fixture used for 4 hours per day at 10¢ per kilowatt-hour saves $1.30 per month in utility costs, and the tubes last longer than seven sets of incandescent bulbs.

Lighted Shelves

Lighted shelves are manufactured. They are excellent above a desk. They contain a fluorescent source and a special lens to avoid glare. They not only light the work surface but also provide handy storage. Mount the shelf 15 to 18 in. (38 to 46 cm) above the desktop and 9 to 12 in. (23 to 30 cm) back from the edge of the desk at the chair.

smart
lighting for
businesses
20

Small businesses often occupy spaces of 300 to 3,000 sq ft (28 to 279 sq m) in large cities and small towns. Some sell products such as clothes, food, drugs, jewelry, flowers, or picture frames. Some provide services such as insurance, real estate, medical, haircutting, travel, or banking. All are competitive. They all want to create a positive impression. Competition heightens the need to have a positive image and to be productive. Images are projected to a great degree by lighting the space, in all cases, and the products,

in some cases. Productivity is affected by lighting. Lighting research in offices and factories has shown that a correlation exists between light levels and productivity. It was found that, within limits, when the light level goes down, so does productivity; when it rises, productivity rises. Other research has shown that employees' morale, motivation, and safety are also influenced by how much light is available. Lighting can have powerful effects.

power for selling

Light is powerful for selling products. It draws customers' attention—the first step to buying. Other senses assist, but without a doubt light makes possible the visual part of the buying decision, and visual appeal is responsible for most impulse sales. Visual appeal induces the customer to want the product. Well-designed store lighting guarantees visibility. It can make products vivid and focus customers' attention. It becomes the silent persuader. Needless to say, if the product is large, visibility is easier. If the color of the product is similar to its background, visibility is hindered and more light is required. If the color of the product is dark, details are less visible. Light permits identification of a product

and reveals its parts. Products are moved frequently; lighting needs to move also.

Products differ, and different kinds of stores require different lighting. For example, clothing stores need soft general light, brilliant displays, and sufficient light focused on important areas such as cashier stations. By contrast, food and drug stores need bright general light, well spread and delivered from direct sources. This type of light reveals colors of products, creates sparkle, and commands a brisk atmosphere. The light sources should be hidden as much as possible in all stores by high ceilings, baffles, or ceiling systems. Recessed trapezoidal ceiling systems obscure lighting fixtures well by

Light a display three times as bright as its surroundings. Adapted with permission from p. 75 of Contract Magazine, December, 1979.

creating a visual barrier from the usual viewing angle (45°).

Rules of Thumb for Commanding Customers' Attention

• Light a display at least three times as bright as its surroundings.

• Sometimes special displays need to be five to ten times as bright.

• Products should visually dominate a commercial space.

• Illuminate display and appraisal areas at similar levels. (Exceedingly different levels complicate buying decisions. Therefore, one area should not be more than three times brighter than another. For example, if sweaters are displayed in a showcase and appraised in a dressing room, the showcase should not be more than three times as bright as the dressing room.)

• Skin tones need to be enhanced, not destroyed, by the light at mirrors. (Customers need to feel good about what they see in the mirror. Consult the details on mirrors in Chapter 9.)

• Illuminate products with the same color of light under which they will be used. (This rule applies not only to clothes but also to household products. For example, in a frame-it-yourself shop, customers coordinate their artwork with the choices of frames and mat colors. These products are chosen in the shop and must look the same when they are hung on the wall at home or in the office. Therefore, light sources in the shop should permit judging how the colors would appear under incandescent light for residences, cool-white fluorescent for offices, and maybe sodium and metal-halide for other commercial spaces.)

Rule of Thumb for Contrast of Visible Areas of Light Within a Space

• Visible areas should not contrast severely with one another—no more than one to five times brighter. (At any one time a customer should not be able to see a lighted area that is more than five times brighter than the dullest area.)

Product Displays

Where are products displayed? Some displays are horizontal; some are vertical. Horizontal displays are on table or countertops, or in showcases. Vertical displays are in racks, wall-hung, or free-standing. (See Chapter 11 to learn how different materials are enhanced by light from different locations.)

HORIZONTAL DISPLAYS

Horizontal displays require downlighting, showcase, or countertop lighting. Downlighting is produced by overhead fixtures. If the display is in a glass showcase, minimize glare by positioning the fixtures above the front edge of the case. Avoid downlights at mirrors, however; light directly above customer's heads makes unattractive shadows and distorts facial features. Downlight is produced usually by R (reflector) and PAR incandescent sources in 50- to 300-watt sizes. Horizontal displays are normally lighted uniformly.

Showcase lighting can be produced by fluorescent or tubular incandescent sources. It should be broad and shadow-free. Since showcases are enclosed, heat builds up. Fluorescent sources are cooler than tubular incandescent. Use them whenever possible. Attract attention to the showcase by having the amount of light three times as bright as adjacent areas. (See Chapter 11 for ways to light furniture.)

Countertop lighting can be direct or diffused. Direct light is intense and focused on a product. It is produced by regular- or low-voltage incandescent sources. Regular voltage requires reflector spots in 25 watts or more, depending on surrounding light. Low voltage requires a fixture with a transformer, like a high-intensity lamp. Either way, the fixtures should be adjustable, because when displays change, the lighting should change. (See Chapter 4.) On the other hand, diffused is soft and spreads light in the area of products. It can be produced by table lamps with the new fluores-

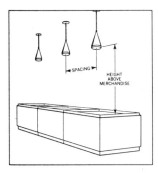

When uniformly spaced PAR and R lamps are used in counter down-lighting, the table at right shows the approximate center-to-center lamp spacing to achieve 50, 100, or 200 footcandles directly under the row.

Lamp Type ①	Height (ft)	50 fc		100 fc		200 fc	
50-watt R-20	4	2 ft	3 in	1 ft	2 in	0 ft	7 in
	6	1	6	0	9	0	4
75-watt R-30 flood	4	2	2	1	1	0	7
	6	1	6	0	9	0	4
75-watt PAR-38 flood	4	(2)		2	0	1	0
	6	3	0	1	6	0	9
	8	2	0	1	0	0	6
150-watt R-40 flood	5	4	3	2	3	1	2
	7	3	4	1	7	0	10
	10	2	3	1	2	0	7
150-watt PAR-38 flood	5	(2)		(2)		2	0
	7	(2)		3	0	1	6
	10	4	0	2	0	1	0
	14	3	0	1	6	0	9
150-watt R-40 spot	7	(3)		(2)		2	1
	10	(2)		3	0	1	4
	14	4	2	2	0	1	0
	20	3	0	1	4	0	8
150-watt PAR-38 spot	7	(2)		(2)		3	5
	10	(2)		3	6	1	8
	14	4	8	2	4	1	2
	20	3	6	1	8	0	9
300-watt PAR-56 wide flood ③	7	(2)		(2)		5	0
	10	(2)		7	6	3	6
	14	10	0	5	0	2	6
	20	7	6	3	6	1	9
300-watt R-40 flood	5	(2)		4	3	2	3
	7	6	6	3	4	1	7
	10	4	3	2	3	1	2
	14	3	4	1	7	0	9

① Data for new, unshielded lamps. Illumination will decrease slightly with age. Louvers, baffles and reflectors will change required spacings.
② Exceeds this level directly beneath lamp at this height.
③ Oriented with wide segment of beam parallel to row.

cent screw-based bulbs of 8 watts or more, or incandescent bulbs of 25 watts or more, depending upon the amount of light needed to attract attention.

VERTICAL DISPLAYS

Vertical displays require accent, valance, cornice, or luminous panel lighting. Accent lighting is confined and intense. It can come from incandescent or high-intensity discharge (HID) sources. On the other hand, light from valances, cornices, and luminous panels can be well spread and less intense.

Accent light is used for feature displays. Normally, it is created by a point source. It should reveal details and attract attention but not detract from the rest of the space.

Rule of Thumb for Accent Light
The center of the light should fall on the important parts of the display at a 60° angle.

This rule ensures the most effective lighting with natural-looking highlights and shadows, and it minimizes glare for customers approaching from the other side.

Accent fixtures should be adjustable. Use recessed, surface-mounted, or track fixtures. Tracks can accept additional fixtures if more

60 degree angle.

light is required. Regular voltage is suitable, but low voltage is preferable. Equip regular-voltage fixtures with R or PAR bulbs. R bulbs put out an area of light with indistinct edges and less light. PAR bulbs put out an area of light with distinct edges and more light. Either way, use a spot bulb for a smaller, brighter area; use a flood bulb for a larger, softer area. Regular-voltage PAR's or R's are available in 50- to 300-watt sizes. Low-voltage PAR's, either spots or floods, are available in 25 to 75 watts. Particularly noteworthy are the new multifaceted reflector types that produce an optically controlled beam for more brilliance and contrast—very good for displays. They are available in 20, 50, and 75 watts. The other low-voltage PAR's are 25, 30, and 50 watts.

Valance and cornice lighting use fluorescent sources. They are excellent for wall displays, clothing racks, and shelves. Install the fluorescent tubes a distance out from the vertical display equal to one-fourth of the distance over which the light is expected to fall. For example, if the display is 4 ft (1.2 m) high, the fluorescent tube should be 1 ft (.3 cm) out from the display. (See cornice and valance lights in Chapter 5.) Parabolic reflectors can spread fluorescent light even farther and make it more energy-efficient.

On the other hand, a valance or cornice board can hide an electric track and incandescent accent fixtures, adding brightness to special features within a wall display. Either fluorescent or incandescent light from vertical displays reflects into the space, giving more general illumination and a spacious feeling.

Luminous panels can be vertical or horizontal. They can produce soft backlighting or underlighting for products. Vertical luminous panels backlight shelves. Horizontal panels underlight merchandise, creating a tabletop or shelf surface. Both use fluorescent sources. Both are good settings.

To build a luminous panel, construct a covered box. If the covering is translucent, separate the tubes by a distance equal to one and a half times the depth of the panel. For example, if the panel is 4 in. (10 cm) deep, put the tubes 6 in. (15 cm) apart. Put the tubes well back from the edges at the top and bottom or at both sides, especially if the covering is transparent. Paint the inside of the box off-white to reflect light. (See luminous panel directions in Chapters 8 and 9.)

Additional light can be on the products from the front to highlight and reveal details, if backlighting is not enough. (See Chapter 11.)

A luminous area can be created by diffusing light through translucent fabric. The luminous area needs to be brighter than the surroundings. Fabric spreads the light like a diffuser.

Diffuse light through fabric. Adapted with permission from p. 75 of *Contract* Magazine, December 1979.

Valance light for vertical displays.

NUMBER OF FOOTCANDLES FOR STORES

	HIGH ACTIVITY		MEDIUM ACTIVITY		LOW ACTIVITY	
Showcase and Wall Displays	100	(1,000 lux)	75	(750 lux)	30	(300 lux)
Feature Displays	500	(5,000 lux)	300	(3,000 lux)	150	(1,500 lux)
Circulation	30	(300 lux)	20	(200 lux)	10	(100 lux)

Amount of Light for Stores

The Illuminating Engineering Society recommends that levels of light required for stores be determined by the amount of activity generated in the area. They classify activity into three categories—high, medium, and low. High activity is easily recognizable merchandise, rapid viewing time, and impulse buying. Medium activity is familiar merchandise, longer viewing time, and thoughtful decisions. Low activity is unfamiliar merchandise, with assistance and time required for viewing and decisions.

Energy-efficient light sources, incandescent, fluorescent, mercury, and high pressure sodium, have been developed over the past few years. Since 60 percent of the energy consumed in retail business is for lighting, it is more important than ever to utilize these more efficient sources. They do not have to decrease the quality of lighting. In fact, they actually can increase it with better color-rendering properties, longer rated life, and higher lumens per watt. Use them. If a fluorescent source is now being used, replace it with the newer reduced-wattage fluorescent. Replace a 150-watt incandescent with a 75- or a 120-watt elliptical reflector. Replace a 500-watt incandescent with a 450-watt self-ballasted mercury; the ballast is in the lamp and not required in the luminaire. Replace a 175-watt mercury with a 150-watt high pressure sodium, and replace a 400-watt mercury with a 325-watt metal halide. Watts and money can be saved.

Displays in Windows

Displays in windows require light. The amount required depends on many things. Are the displays viewed during the day or at night? Are they viewed in competition with other windows? Do the windows receive reflections from other light sources, such as streetlights?

Rule of Thumb for Lighting Display Windows
Feature displays require high illumination during the day when the outside light is bright and at night in a highly competitive area when external lights can reflect.

A photosensor and a dimmer can raise the footcandle level in the window when daylight gets brighter, and they can lower the level at night when the higher level is not needed. Consequently, these devices can make sure that the window is bright enough to have the potential of attracting attention without wasting electricity.

The Illuminating Engineering Society divides window lighting into general or feature displays, and they recommend these amounts:

NUMBER OF FOOTCANDLES FOR STORE WINDOWS

	DAYTIME	
General Display	200 (2,000 lux)	
Feature Display	1,000 (10,000 lux)	
	NIGHTTIME	
	Highly Competitive	Noncompetitive
General Display	200 (2,000 lux)	100 (1,000 lux)
Feature Display	1,500 (15,000 lux)	500 (5,000 lux)

Some display windows have no back wall, and anyone can see through them into the store. They are called see-through displays. External and internal brightness affects the amount of light required in the see-through display. External light (daylight, mall lighting, or streetlighting) enters these windows, adding to or detracting from the window display. Likewise, internal light affects the attention-getting quality of the see-through display. Consequently, the brighter the interior, the brighter the display needs to be. See-through displays also permit outside light to enter the store, adding to or distracting from the interior brightness. They can reduce the store's dependence on electric light, but they require careful planning to be balanced, glare-free, and successful.

Tables adapted from *IES Lighting Handbook, 1981 Application Volume.*

Features in a display window require higher illumination.

A see-through display window permits outside light to enter.

combining daylight and electric light

Businesses of all kinds utilize daylight to reduce their dependence on electricity. Photosensors can monitor the amount of daylight and adjust the electric lighting to provide constant illumination levels at desks or in shops. The payback time (the time required to save enough to equal installation costs) for incandescent lights is 1 to 2 years; and for fluorescent lights it is 2 to 5 years. In an incandescent system, photosensors create long-term savings with less electricity for lighting and air conditioning and with less bulb replacements. In a fluorescent system, photosensors create savings with less electricity for lighting and air conditioning only, and bulb life is not affected. Dimming incandescent light makes it more golden; dimming does not affect the color of fluorescent light. Photosensor systems operate automatically to dim or to raise light levels. They can be installed by owners or renters in new or existing structures.

Sample Electric Cost
(Reprinted by permission from Lutron.)
A photosensor could save $484 in electricity at 6¢ per kilowatt-hour in an office lighted by daylight and 72 ceiling 40-watt fluorescent bulbs operating for 4,000 hours per year. With an installation cost of $1,775, the payback time would be 3.7 years.

A photosensor could save $678 in electricity at 6¢ per kilowatt-hour in a shop lighted by daylight and 33 ceiling 100-watt incandescent bulbs operating for 4,000 hours per year. With an installation cost of $800, the payback time would be 1.2 years.

power to create atmosphere

Low-Voltage Tube Lights

Light can create atmosphere in many ways; it can create visual excitement. For instance, tiny lights create a twinkling, starlike atmosphere. Tiny lights can be contained in tubes, whether flexible or rigid. These are low-voltage tube lights. The bulbs are ¼ watt each and are positioned 2 to 20 in. (5 to 51 cm) apart inside the tube. The closer they are together, the greater the sparkle. The bulbs are distinctive and easily maintained, since they last for 50 years. They require low-voltage transformers to reduce the normal current to 12 volts. They can be made to chase (blink on and off), as in discos if such an effect is desired. The tubes can be hung as a chandelier or a curtain, or used to outline architectural edges. They can create an atmosphere in any space, whether commercial or residential.

Sample Electric Cost
In a large hotel lobby, a curtain of 9,000 of these bulbs costs 23¢ per hour to operate at 10¢ per kilowatt-hour.

Light can differentiate areas within the business. For instance, public areas can be brighter and appear more active, while private areas can be more subdued with brighter accents for focal points. Also, light can establish moods, such as in restaurants.

Restaurant Lighting

Restaurants rely heavily upon light to create atmosphere. A brisk-paced atmosphere is created in fast-food restaurants by bright light. Leisurely dining is created by subdued lighting to keep a slow pace. Create an atmosphere according to the type, the standard of accommodation, and the potential trade of a restaurant. Although atmospheres may differ, lighting for essentials does not. The essentials are tables, serving counters, work stations, and public facilities. Light is needed at these places. Lighting design for essentials can be different. Needless to say, the more time and money available, the more opportunities there are for designing powerful lighting beyond the essentials. Nonetheless, essential lighting can be designed effectively.

Restaurant Lighting Guidelines
- Large sources of light (large luminous ceilings, ceiling panels, or large fixtures) need to yield soft, not harsh light.
- Low-level lighting slows down the eating pace.
- Fixtures hanging below the ceiling (chandeliers or pendants) should relate to tables or other furniture. Conversely, fixtures on the ceiling can relate to the space or its divisions.
- Intense downlights should not be over customers' heads. Harsh shadows are created under eyes and noses.
- Tabletop candles alone do not yield enough light for a whole restaurant.
- Pools of light on tabletops next to darkness create an intimate atmosphere.
- Direct light (downlight in pendants or recessed fixtures) on tabletops creates sparkle on the tabletop items.
- Bare, low-wattage bulbs (in chandeliers, ceiling fixtures, wall fixtures, or flexible tubes) create sparkle. None should be over 15 watts.
- Wall-hung fixtures add to the decoration but do not produce sufficient light for most restaurants.
- Wall-washing emphasizes the wall finish and also reflects light.
- If low illumination is the design goal, then higher illumination should be available for cleaning the restaurant.
- Preset dimmers can change lighting levels for different atmospheres without complicated settings; for example, moderate level for a banquet, brighter with a focus on the after-dinner speaker, and subdued for dancing.

DESIGN CONSIDERATIONS

Consider the time of day a restaurant will be used. If it is to be used during the daytime and has no windows, the illumination level must be higher than at night. The contrast of light between the inside and outside should not be too great—one to five times is considered a reasonable ratio. Consequently, during the day when the outdoor illumination level is high, the interior level must also be relatively high; otherwise eye adaptation could be momentarily painful. (Remember the last time you came out of an afternoon movie on a sunny day and how much your eyes hurt for a moment or two?) Photosensors are available to automatically control the light inside a restaurant on the basis of the level outside and to maintain a comfortable contrast.

Consider the various uses of the restaurant space. For example, it may be rented for birthday parties, club banquets, business meetings,

Large sources of light need to be soft.

Pools of light next to darkness create an intimate atmosphere.

Wall-washing reflects light.

and wedding receptions. Each function requires a different atmosphere. Preset or programmable dimmers can create whatever atmosphere is required with no guesswork. Once programmed, they operate with a simple on-off switch.

Consider the volume of the space and the effect desired when choosing fixtures. Normally, larger volumes of space accept larger visible fixtures. Smaller volumes accept small fixtures. However, hidden fixtures (recessed, cornice, cove, valance, luminous ceiling or panel) are good for both small and large spaces. Combine the style of fixtures with the design of the interior, arrangement of the space, and structural constraints for the desired effects. Stylized fixtures can create an atmosphere themselves. Overall, visible fixtures should be pleasing to look at (lighted or unlighted), mechanically operable, easily maintained, and long-lasting. The range of effects are broad and hard to describe. In fact, professional illustrators hate to draw nighttime restaurant interiors because the atmosphere created is so difficult to reproduce graphically. For lighting designers, the process of creating atmosphere cannot be prescribed. Suggestions throughout this book will assist in designing restaurant lighting.

In designing, be sure to consider the contrast between the kitchen and the dining space. If the light in the dining space is subdued, make sure that bright kitchen light does not blast into the dining area. Otherwise, the artfully created atmosphere for the customers will be destroyed. Likewise, it will be too much contrast for waiters' and waitresses' eyes.

businesses with color-matching tasks

Many businesses are based upon seeing or matching colors correctly—medical testing labs, printers, hairdressing salons, fabric shops, wallpaper stores, clothing stores, and others. Sometimes matching colors is critical (medical); sometimes it is less critical, but important, nonetheless (printing). Special fluorescent bulbs are made for critical color recognition. They are called color-matching fluorescent. They are more expensive than cool-white fluorescent. Their light is the color of daylight and they are the best lighting source for color detection. In businesses in which color matching or judging is less than critical, the lighting industry recommends using a combination of fluorescent and incandescent lights. For example, in a hairdressing salon, use all warm-white fluorescent lights, or use incandescent lights at mirrors and color-matching fluorescent lights for general illumination so that hair color can be judged adequately. Warm-white fluorescent lights or incandescent lights flatter any color of skin and all natural hair color. However, these sources alone would not be sufficient to judge shades and tones for coloring hair. On the other hand, using cool-white fluorescent only would reduce the red tones in the hair, face, and clothes. It would be unflattering and would create surprises when the customers get home. Always use color-matching fluorescent bulbs where detection of color is critical.

power for productivity

In both shops and offices, lighting has the power to create a productive work environment. When the light level is low, visual efficiency is poor and work can slow down. The cost of providing the recommended light levels is typically only 1 percent of employees' salaries. Good light is good for business. Businesses that provide a service evaluate their light by its effect on people and their tasks. Illuminate the task first and the space second. Sometimes a task fills a space. In such cases, task lighting needs to be uniform. The amount of light required for a task depends upon how difficult the the task is, how important it is, and the ability of the employee to see. The more difficult the visual task, the more important the task, and the older the employee, the more light is required. The requirement for reading ranges from easy at 30 footcandles to difficult at 150 footcandles. (Consult the *IES Lighting Handbook 1981 Application Volume*.) Tasks, also, need glare-free and shadow-free light. Consequently, task light must not come from above or in front of the task. (See Chapter 19.)

Uniform and Nonuniform Lighting Design

Design lighting for small commercial spaces by deciding whether the whole space must be uniformly lighted or not. Uniform lighting spreads the same amount of light from wall to wall. Why light the whole office at 50 footcandles when much of the light falls on the floor? Uniform lighting is not always the best. Nonuniform lighting makes some areas brighter than others, usually determined by task requirements or desired effects. Nonuniform lighting can provide 50 footcandles at the desk and spread 10 footcandles elsewhere. This practice can be more energy-efficient, more interesting, and more adaptable.

SOURCES

Sources for uniform lighting are either direct or

indirect. Direct light is usually produced by fluorescent sources in average-height spaces, and metal-halide, sodium, and deluxe mercury sources in high-ceilinged spaces. Careful planning is required to get smooth light throughout the space with no hot spots (areas of bright light) anywhere, including the ceiling. Smoothness is determined by fixture design, placement, and bulb wattage. If the fixtures are well engineered to put the light down without brightness at the source, correctly positioned, and correctly lamped (have the correct bulb and type wattage), the light will be distributed uniformly.

Indirect light is produced by either metal-halide, deluxe mercury, sodium, or fluorescent sources. Metal-halide, mercury, and sodium cost more to purchase and install than fluorescent, but use less electricity. Therefore, the initial cost is offset by lower long-range cost. Indirect light is bounced from somewhere else (the ceiling or a wall) before it falls into the space. Indirect light can supply illumination for either uniform or nonuniform lighting.

Nonuniform lighting sources can be all direct or both direct and indirect. Nonuniform direct light is produced by point sources (regular or low-voltage incandescent and high-intensity discharge). Indirect light can be produced by point sources (high-intensity discharge) or linear sources (fluorescent). When used as direct light, point sources create highlights and shadows; when used as indirect light, they can spread illumination over a broad area. Linear sources do not create highlights and shadows however they are used.

FIXTURES

All fixtures for direct light are positioned to put light down. Direct lighting fixtures can be on the ceiling, recessed, surface-mounted, or suspended. On the other hand, indirect lighting fixtures are positioned to put light on a large architectural surface—a ceiling or wall. They can be ceiling-hung, wall-mounted, furniture-integrated, or free-standing. Be aware that furniture-integrated and free-standing indirect fixtures are classified as furniture for tax purposes. Therefore, they can be deducted at the accelerated rate of furniture rather than as building equipment. Likewise, such fixtures do not require disturbing the fire-resistant ceiling of a structure, and a lower fire insurance rate is possible. Many considerations are required to decide which system is the best. They include aesthetics, cost of installation, tax break, insurance rate, cost of maintenance, cost of electricity over time, and available funds. The decision is complex at best, and sometimes specifications for bids are required. Consult lighting designers and manufacturers to help with these decisions and specifications.

LIGHT LEVELS

In uniform lighting, the whole space is illuminated to the level required for the task. In nonuniform lighting, the task area is illuminated to the level of the task, and the general space (ambient light) is lighted at a lesser amount. The ambient light should equal one-third to one-fifth the amount of light for the task. For example, if a drafting table is lighted to 200 footcandles, then ambient light can be 40 footcandles. The whole room is not lighted to the same level, and less energy is consumed. Ambient light can be aesthetically pleasing and define the space. It can provide safe passage and enhance the image of a business. Some areas of ambient light can be brighter than the task area. They can be up to 10 times brighter. Bright ambient light can come from a large surface washed with light—wall-washing, cove lighting, and uplighting. Soft ambient lighting can come from accent, downlighting, and art lighting.

OPEN-OFFICE LIGHTING

Open offices (offices with furniture-hung partitions and no walls) present lighting problems, as reported in an analysis of five open-office lighting systems by Noel Florence (see the bibliography). Open offices have partitions and furniture that can be changed, usually requiring repositioning of light fixtures. Lighting for open offices can come from ceiling-hung sources (direct) yielding uniform illumination, from task-located positions yielding nonuniform illumination, from widely spaced ceiling sources and furniture (undercabinet or shelf-mounted) sources (both direct), or from furniture-mounted or free-standing uplight sources (indirect) for the space and furniture-mounted (direct) for the task. Shelves and cabinets overhanging furniture can block the light from the ceiling. Also, direct light suspended from cabinets and shelves for tasks can cause reflected glare and poor visibility. Noel points out that office lighting systems should be evaluated for energy effectiveness, which includes energy input, visual comfort, and relative visibility at the task location. He states that spacing of the task locations also affects the lighting effectiveness. Therefore, a single measure of watts per square foot or square meter (building watts) is not the only appropriate measure. Likewise, watts per task location (people watts) should be considered. These two measures should be used to design the best system for open-office lighting.

Fixture Determination

In commercial spaces, the type of ceiling determines the options for ceiling fixture installation. Ceiling construction is usually one of the following types—wood ceiling joists, suspended grids, open bar joist, or concrete and steel. Wood ceiling joists, suspended grids and bar joist types accept recessed or surface-mounted fixtures. If the ceiling is concrete and steel, the fixtures must be surface-mounted, unless they are set in before the concrete is poured.

The decision of what fixtures to buy depends on many decisions and trade-offs—color of light, initial cost, operating cost, ease of maintenance, potential noise (from ballasts), payback time, and aesthetics. Choose wisely; well-engineered fixtures pay you back over time. After determining which fixtures are possible, develop the reflected ceiling plan. Be sure to pay attention to the pattern the fixtures make on the ceiling. They are unmistakably in view, and lighting design for offices is composed of more than fixture determinations.

Maintenance

Providing sufficient light is only the beginning.

Maintenance is required and includes bulb replacements, fixture adjustments, and cleaning. Many an appropriate and even spectacular lighting design has been ruined by improper bulb replacements or lack of fixture adjustment. (How often have you sat in an expensive hotel meeting room and could not see the speaker because the ceiling fixtures—the ones with bulbs—were facing every which way but the direction intended—on the speaker?)

Poor maintenance does not happen just in hotels; it happens in offices, stores, and even homes. In hotels and offices, everyone waits for the maintenance staff to fix the fixtures. In stores, poorly lighted displays should cause the owner to pay attention. In homes, fixture maintenance is usually thought of as bulb replacement only. But, any fixture requires cleaning. Many installations providing ample light have been reduced to insufficient light if the fixtures have not been cleaned. Dust must be removed. It can cause a 20 percent loss of light, impeding the necessary work, particularly if the initial level was minimal to conserve energy.

To further complicate matters, a light bulb produces less light as it ages—12 to 15 percent less. Consequently, the amount of light is decreasing all the time until the fixture is re-lamped (new bulbs put in) and recleaned.

Maintenance means organizing the maintenance staff, training and supervising them to relamp, adjust, and clean the fixtures. It also means cleaning, repainting, or rewallpapering the walls and ceiling to reflect the greatest amount of light.

power of glare

Light has the power to cause glare, which defeats the purpose of lighting. Glare can be reflected or direct.

Reflected Glare

Glare can reflect from printed material (glossy magazines and glossy print). It is called veiling reflections, because they veil or obscure what you are looking at. In commercial spaces, glare can reflect from glass showcases, transparent packages, large windows, mirrored walls, or other shiny surfaces. Reflected glare blurs visual information, is annoying, and gives inaccu-

rate visual cues. For example, in a rehabilitated building, large round support columns were painted a glossy red. Unfortunately, poorly designed lighting caused an hourglass-shaped glare on the columns which made them appear bent rather than perpendicular, and they did not appear architecturally firm.

Sometimes, mirrors, marble, or other shiny surfaces reflect an image of the space, which looks like additional space. Reflection confuses people. For example, an interior design instructor related how he designed a restaurant and covered large square columns with mirrors, hoping to hide them. He did. However, because of the reflection, people bumped into them. He was called back to redesign. Be careful of both reflections and glare.

Direct Glare

Glare can come directly from fixtures. This type of glare is a concentration of light. It draws your attention and can be fatiguing for the eyes. Therefore, ceiling fixtures should not be glaring, particularly fixtures over 4 sq ft (.037 sq m). Manufacturers publish visual comfort ratings for their fixtures—called Visual Comfort Probability (VCP). The ratings are an evaluation of direct glare in a room with a lighting system using a specific fixture and judged by people seated at a specified location. The rating is the percentage of people who describe the system as comfortable.

Rule of Thumb for Visual Comfort

The VCP rating for a lighting system using a specific fixture should be 70 or more.

However, if your space is less than 20 ft (6 m) in both directions, the ceiling fixtures are essentially not in view, and the VCP does not apply.

FLUORESCENT FIXTURES

The visual comfort of ceiling fluorescent fixtures is controlled for the most part by shielding materials—louvers, lenses, and diffusers. Louvers conceal the light bulb from a horizontal to 45° viewing angle, depending upon the shielding ability. They are made of open grids, plastic, or metal.

Lenses and diffusers redirect the light. Lenses actually bend or refract the rays. Diffusers generally spread the light. They are made of either glass or plastic. One type of lens redirects light in two directions, avoiding reflected glare (veiling reflections) that obscures printed materials; it is called a bilateral or batwing lens. A prismatic lens redirects light downward to reduce direct glare at the fixture. Most lenses are made of acrylic or styrene plastic. The price is less for styrene, but it turns yellow and gives a yellow glow to the light. (Nail polish remover will dissolve styrene; use it to test the type of lens if you are uncertain which type you have.) Diffusers obscure the bulb and give a more uniform brightness. They are also usually made of opal glass.

Shielding materials have varying degrees of comfort probabilities. Parabolic wedge louvers are considered the most comfortable (99 VCP). Other louvers and lenses are considered a little less comfortable (50 to 90 VCP). Diffusers are considered the least comfortable (40 to 50 VCP).

Choosing shielding material can be complex. Besides comfort rating, other qualities need consideration—stability of color, cleanability, fire rating, manufactured size compatibility with the job, cost, and light transmission quality.

INCANDESCENT AND HIGH-INTENSITY DISCHARGE FIXTURES

Visual comfort from point-source (incandescent and high-intensity discharge) fixtures is controlled by the design of the fixture. Well-engineered fixtures control glare better and also produce more light—sometimes as much as 400 lumens more per fixture. Some fixtures have shielding materials (lenses and diffusers). Some are rated for visual comfort. (See Chapter 6 for greater details about incandescent fixtures.)

power of heat

Light sources radiate heat. Heat can be uncomfortable for customers and employees. Often, it needs to be air-conditioned out of the space. Moreover, the heat has an effect on some products, causing perishing and fading.

Perishing

Perishing includes both melting and discoloring. Some products melt (candy). Some products discolor (meat, both fresh and frozen).

Perishing can occur at any time. Sometimes discoloration occurs in hours; sometimes in days. Discoloration of food is caused by bacterial growth in the warmth of the light. Both melting and discoloration reduce sales potential.

Fading

Some materials fade. Fading usually occurs between 50,000 and 70,000 footcandle hours (footcandles multiplied by time). For example, if a

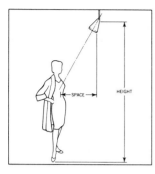

This table shows appropriate spacing relationships and resultant highlight illumination · for spotlighting vertical displays. Data are based on aiming each lamp with the beam axis striking the display about 5 feet above the floor, from a 60-degree angle. Highlight illumination values are those obtained over a small area in the center of the light beam on a surface perpendicular to the beam axis.

Lamp Type	Height (ft)	Space (ft.	in.)	To Obtain These Average Max. Footcandles*
50-watt R-20	8 10	1 2	9 11	50 20
75-watt R-30 spot	8 10	1 2	9 11	155 55
150-watt R-40 spot	8 10 12	1 2 4	9 11 0	620 220 115
150-watt PAR-38 spot	8 10 12 14	1 2 4 5	9 11 0 2	760 270 140 85
200-watt PAR-46 narrow spot	10 12 14 16	2 4 5 6	11 0 2 4	1080 550 335 225
200-watt PAR-46 medium flood	10 12 14 16	2 4 5 6	11 0 2 4	325 160 95 65
300-watt PAR-56 narrow spot	12 14 16 18	4 5 6 7	0 2 4 6	1070 650 435 310
300-watt PAR-56 medium flood	12 14 16 18	4 5 6 7	0 2 4 6	370 220 150 105

*Based on new, unshielded lamps. Louvers or baffles may reduce these values slightly.

Reprinted by permission of General Electric Company.

product with unstable dyes is displayed at 500 footcandles for 100 hours, it can fade. Pale colors often fade faster than more brilliant colors. Fading is more apparent when some parts fade and some do not because they are away from the light. For example, neckties folded in display cases can fade on the edges and on the top. Products should be rotated out of the light every 10 days to control fading. Fading is less apparent if the whole product is uniformly exposed to the light. Products should be spread, not stacked, in and on cases. All light sources cause fading. The hotter the source, the more the fading. Keep light bulbs at least 9 in. (23 cm) away from products.

Cooler Light Sources

In most cases, the lower the wattage, the lower the heat. Fluorescent and low-voltage incandescent sources are low and cool, particularly the new 20-watt, low-voltage PAR's. However, a special regular-voltage incandescent source has a cool light beam. Use it in installations where greater light output is required. It is a regular-voltage, heat-reducing PAR bulb. It filters out about two-thirds of the heat, sending it out the sides rather than with the light. Use it to either reduce heat or to increase the amount of light for the same amount of heat. For example, when competition is keen, the light in a window display can be increased to attract attention, and the potential of fading can be kept the same with a cooler light source in a higher wattage. Likewise, when displaying food, the amount of light can be kept the same and the radiant heat reduced with a cooler light source.

power for steering

Light has the power to steer people in the direction you want them to go. Bright enough light attracts their attention and permits them to recognize what they are looking for—the exit, the public facilities, the cashier station. Make the areas you want to be seen three times brighter than the surrounding area and reduce questions about directions.

Cold Cathode Lighting

For many years, cold cathode (or neon) lights have been used as outdoor signs to steer people to places of business. They are unexcelled in brilliance and ability to be molded into any shape. Because of their attention-getting qualities, they are useful for accentuating interior architectural features and enhancing commercial spaces. Because of their ability to be molded, they have been gaining popularity in residential spaces as artful sculpture and outlines for interior architecture. However, in any space, cold cathode can be overpowering, noisy, and too functional-looking. Great skill is required in using such lighting.

Emergency Lighting

In business spaces, emergency lighting is required for a minimum of 90 minutes to comply with codes. Some emergency fixtures are designed to use auxiliary electrical power; some use batteries. Some are functional-looking units; some are architecturally integrated. Some are simply energy-producing packs to fit into the existing fixtures. Choose the most aesthetic type to suit your pocketbook. Be careful where the emergency lighting points. Emergency lighting can confuse people, especially if it

Cold cathode light emphasizes the architecture with brilliance.

glares in their eyes. It should be designed to get people to the exits in fire or dense smoke.

vocabulary
for
professionals

21

Professionals need to be aware that the language of lighting is confusing. Sometimes two terms mean the same thing; for example, illumination and illuminance both mean "amount of light." Sometimes a term means two different things; for example, skylight means a "window on the roof" or "light from the sky." Further, terms that describe measurements of light based upon feet are slowly being changing to terms based on meters.

The communication process is further impeded because building terms differ regionally. A recessed electrical connection can be called a junction box, an outlet box, an electric box, a j-box, a tangle box, or a splice box. A baseboard electrical connection can be called a baseboard receptacle, a duplex outlet, or a wall plug.

In addition, many people design lighting—lighting consultants, interior designers, architects, electrical engineers, manufacturers' reps, contractors, landscape architects, electricians, and lighting showroom personnel. Indirectly, so do magazine editors, educators, and maintenance staff. These people are not trained the same way and do not talk the same way. However, they should speak the language of the lighting profession.

Professional language is confusing because it often conflicts with the language used by the general public. For example, when a professional talks about a lamp, the general public calls it a light bulb. Within the profession, a lamp manufacturer could make light bulbs or portable luminaires—depending upon which vocabulary is being used.

To facilitate communication, use the professional language with other professionals and use the everyday terms with the general public and lighting clients, unless they are also lighting professionals. Use the following vocabulary list to equate professional terms and everyday terms.

Professional Terms	Everyday Terms
Accent lighting	Spot lighting
Ambient light	General illumination
Controls	Switches, dimmers
Fenestration	Windows
Fenestration on the roof	Skylight
Ground light	Light reflected from the ground
Hard wiring	Built-in wiring

Housing	Part of the fixture surrounding the bulb
Illuminance	Light (meaning that which is produced)
Illuminance	Illumination
Lamp	Light bulb
Light source	Type of light produced— incandescent, etc.
Linear source	Fluorescent or tubular incandescent
Luminaire	Fixture
Luminaire	Light (meaning an object producing light)
Nonuniform lighting	Different amounts in the space
PAR lamp	Floodlight
Point source	Incandescent, mercury, sodium, metal-halide
Portable luminaire	Table or floor lamp
Reflected and direct glare	Glare
Retrofitting	Putting a different light source in an existing fixture
Skylight	Light reflected from sky and clouds
Solar radiation	Sunlight
Task light	Worklight
Uniform lighting	Same amount of light throughout a space

professional language clarification

Terminology within the profession is highly technical. Definitions are precise, qualifying, and scientific. Most defy memorization. The beginning professional can grasp the technology better with simple definitions. Therefore, the following list clarifies professional language without severely sacrificing technical quality.

Measurement Terms

Aiming Angle: Angle from which the light falls on the surface lighted. (In this book, 0° aiming angle is perpendicular to the surface being lighted.)

Candlepower: The intensity of light from a source in a specific direction. (Lamps are tested and the candlepower distribution is indicated in a diagram or chart. Both illustrate light distribution, in quantity and in direction—very useful for careful, calculated lighting design.) The measurement is in candelas.

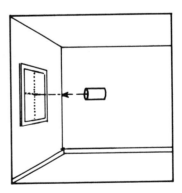

0 degree aiming angle.

Aiming angle is shown at (1).

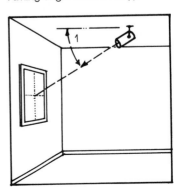

Chromaticity: Degree of warmth or coolness of the color of the light from a particular source.

Coefficient of Utilization: Lumens reaching a surface divided by lumens produced by the lamp. (The amount received is affected by the surface's distance from the lamp and the reflectance of that surface.)

Color-rendering Index: Color-rendering ability of a particular light source, rated between 1 and 100. The higher the number, the truer the color of the object. (For example, cool-white fluorescent is 66; deluxe cool-white is 89.)

Equivalent Sphere Illumination (ESI): A measure of illumination quality, rather than just quantity. It takes into account the reduction of visibility caused by veiling reflections.

Footcandle (FC): A measure of light produced by one candle uniformly onto a surface 1 sq ft in area from 1 ft away. Translate footcandles to lux approximately by multiplying footcandles times 10.

$$\text{footcandles} = \frac{\text{lumens}}{\text{square feet}} \qquad \text{lux} = \frac{\text{lumens}}{\text{square meters}}$$

Footlambert (FL): Measure of reflected or transmitted light intensity coming from a surface. Footcandle times reflectance ability of a surface equals footlambert. For example, a white ceiling with 80 percent reflectance and 100 footcandles falling on it would reflect 80 footlamberts of light. (We see footlamberts not footcandles, because we see light reflected from something.)

Inverse Square Law: The amount of light decreases by the square of the distance it travels.

Lamp Depreciation: The lumens produced by a light source decrease over time. Consequently, the middle amount of lumens over the life of the lamp is less than the initial amount. Manufacturers indicate this amount as "mean" lumens.

LR%: Light reflectance percentage or average amount of light reflected from a particular color surface.

Lumen: A measure of the light produced by a light source. (Manufacturers specify lumens, which can be expected initially from their light sources.)

Luminaire Efficiency: Lumens emitted by a luminaire divided by lumens generated by the lamp.

Luminous Efficiency: Lumens produced by the lamp divided by watts required for the lamp.

Lux: Metric measure of light produced by one candle uniformly onto a surface 1 sq m in area from 1 m away.

Veiling Reflections: Glare reflected in the visual task, which reduces contrast and therefore visibility.

Visual Comfort Probability (VCP): An evaluation of direct glare in a room with a lighting system using a specific fixture, as judged by people seated at a specified location. The rating is the percentage of people who describe the system as comfortable. (The higher the VCP, the better.)

Lighting System Terms

Initial Cost: Cost of equipment, lamps, and installation.

Kilowatt Hour: Measure of electrical energy consumed; 1 kilowatt is equal to 1000 watts for one hour.

Low-Voltage Lighting: Luminaires and lamps that operate at 5.5 or 12 volts, in most cases, not at the standard 120 volts.

Mortality Curve: A graph of the rate and number of lamps expected to fail before their rated life. (These curves assist in determining the best uniform relamping time.)

Operating Cost: Cost of electricity, cleaning, and relamping.

Payback Time: Time required to save enough in operating costs to pay back the initial cost of a particular lighting system.

Rated Life: Number of hours at which one-half of the lamps of any one type are expected to fail.

Uniform Relamping: Relamping each luminaire in a commercial installation with new lamps, regardless of whether or not they have failed.

Lighting Fixture Terms

Ballast: Electrical device that supplies the proper current to start and operate a lamp. (Required by fluorescent and high-intensity discharge lamps.)

Prewired Recessed Fixture: Conduit, connector, and junction box attached. (Saves installation time but prohibited by some local building codes.)

Standard Wired Recessed Fixture: Conduit and connector attached without junction box.

calculation methods

In order to gain specific lighting effects such as attracting people's attention, permitting work to be accomplished easily, or balancing other light, professionals need to determine beforehand how to do it. First they must determine the number of footcandles needed. Then, they must calculate the number, position, and size of the light sources. Use any of the three calculation methods—manufacturers' charts, point-by-point, or zonal cavity methods. Manufacturers' charts are suitable for many installations, but not all. Sometimes calculations need to be based on information not contained in the charts, or sometimes more precise calculations are required. Point-by-point and zonal cavity calculations are precise and can be used for any installation.

Manufacturers' Charts

Manufacturers' charts are in their technical catalogs. They are referred to as photometrics. They are in the form of lighting performance data, of candlepower distribution curves, and of quick calculation charts. In addition, manufacturers publish a spacing ratio for their luminaires, which is convenient for designing uniform lighting.

Lighting performance data are published in several different formats. The data compare wattages, distances and spacing for mounting, luminaires, beam spread, and footcandle levels at various angles. The data assists in determining which lamp to use, where to put luminaires, how far apart, and what intensity and spread of light can be expected.

Lighting performance data for wall-washing.
Reprinted by permission of Lightolier.

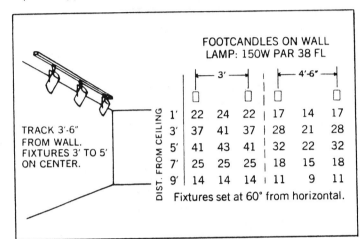

FOOTCANDLES ON WALL
LAMP: 150W PAR 38 FL

DIST. FROM CEILING	← 3' →			← 4'-6" →		
1'	22	24	22	17	14	17
3'	37	41	37	28	21	28
5'	41	43	41	32	22	32
7'	25	25	25	18	15	18
9'	14	14	14	11	9	11

TRACK 3'-6" FROM WALL. FIXTURES 3' TO 5' ON CENTER.

Fixtures set at 60° from horizontal.

LIGHTING PERFORMANCE

ILLUMINATION ON HORIZONTAL PLANE
AIMING ANGLE (A) 30° FROM VERTICAL

LAMP TYPE		75W R-30 FL	75W R-30 SP	75W PAR-38 FL	75W PAR-38 SP	150W PAR-38 FL	150W PAR-38 SP	150W R-40 FL	150W R-40 SP
5'	CENTER TO CENTER (S)	6.6'	2.3'	4.0	1.7'	3.8'	2.0'	7.2'	1.7'
	F.C.	10	40	45	117	104	299	27	140
	BEAM LENGTH (L)	11.0'	5.5'	5.5'	3.3'	5.5'	3.3'	10.0'	4.5'
6'	CENTER TO CENTER (S)	8.0'	2.8'	4.8'	2.1'	4.5'	2.4'	8.7'	2.1'
	F.C.	7	28	32	81	72	208	19	98
	BEAM LENGTH (L)	13.2'	6.5'	6.5'	4.0'	6.5'	4.0'	12.0'	5.4'
7'	CENTER TO CENTER (S)	9.3'	3.2'	5.6'	2.4'	5.3'	2.8'	10.1'	2.4'
	F.C.	5	20	23	60	53	153	14	72
	BEAM LENGTH (L)	15.4'	7.5'	7.5'	4.5'	7.5'	4.5'	14.0'	6.3'
8'	CENTER TO CENTER (S)	10.6'	3.7'	6.5'	2.8'	6.0'	3.2'	11.5'	2.8'
	F.C.	4	16	18	46	41	117	11	55
	BEAM LENGTH (L)	17.6'	9.0'	9.0'	5.2'	9.0'	5.2'	16.0'	7.2'
9'	CENTER TO CENTER (S)	11.9'	4.2'	7.3'	3.1'	6.8'	3.6'	13.0'	3.1'
	F.C.	3	12	14	36	32	93	8	43
	BEAM LENGTH (L)	19.8'	9.7'	9.7'	6.0'	9.7'	6.0'	18.0'	8.1'

MOUNTING DISTANCE (D) IN FEET

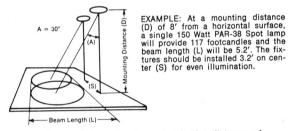

A = 30°
Mounting Distance (D)
(A)
(S)
Beam Length (L)

EXAMPLE: At a mounting distance (D) of 8' from a horizontal surface, a single 150 Watt PAR-38 Spot lamp will provide 117 footcandles and the beam length (L) will be 5.2'. The fixtures should be installed 3.2' on center (S) for even illumination.

In the data at left, center to center is the distance of fixtures in feet for even illumination; footcandles shown are the average in center of beam; beam length is to a point where illumination drops to approximately 10% of maximum or where the illumination becomes insignificant.

Lighting performance data for uniform illumination on a horizontal surface.
Reprinted by permission of Lightolier.

Narrow Beam Spread Lamps	0° Aiming Angle					30° Aiming Angle					45° Aiming Angle					60° Aiming Angle				
Distance (D)	4'	6'	8'	10'	12'	3'	4'	6'	8'	10'	2'	3'	4'	6'	8'	2'	3'	4'	5'	6'
Beam Width (W)	2.1'	3.2'	4.3'	5.4'	6.4'	1.9'	2.5'	3.7'	5.0'	6.2'	1.5'	2.3'	3.0'	4.5'	6.1'	2.1'	3.2'	4.3'	5.4'	6.4'
Spacing (S)	1.2'	1.8'	2.4'	3.0'	3.6'	1.0'	1.4'	2.1'	2.8'	3.5'	0.8'	1.3'	1.7'	2.5'	3.4'	1.2'	1.8'	2.4'	3.0'	3.6'
FC 75W PAR-38 SP	238	106	59	38	26	274	154	69	39	25	336	149	84	37	21	119	53	30	19	13
150W PAR-38 SP	688	306	172	110	76	794	446	198	112	71	972	432	243	108	61	344	152	86	55	38
150W R-40 SP	438	194	109	70	49	505	284	126	71	45	619	275	155	69	39	219	97	55	35	24

Lighting performance data for accent lighting.
Reprinted by permission of Lightolier.

Candlepower distribution curves indicate beam diameter, direction, and footcandle level. The curve is a cross-section and shows the strength of the light anywhere within the beam. Do not be fooled by comparing the sizes of cross-sections without noting the candlepower strength. The cross-sections are not equal in scale.

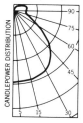

Candlepower distribution curve.
Reprinted by permission of Lightolier.

Quick calculation charts can be used for determining footcandle level for a specified spacing or determining the spacing for a specified footcandle. (They can also be used in zonal cavity calculations, described later in this chapter.)

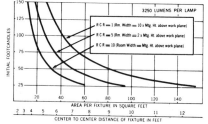

Quick calculation chart.
Reprinted by permission of Lightolier

Spacing ratios can be used to determine the maximum spacing of a selected luminaire in order to get uniform spread of light. Multiply the spacing ratio times the height above the surface to be lighted. For example, when evenly lighting a 8-ft (2.4-m) hallway (the surface is the floor) and when using a luminaire with a spacing ratio of 1, the spacing would be luminaires 8 ft (2.4 m) apart.

For uniform lighting, use spacing ratios, lighting performance data, and quick calculation charts. For nonuniform lighting, use lighting performance data and candlepower distribution curves. These technical aids determine the wattage, the distance, and the angle for a specified footcandle level in order to effectively light an object, a wall, or a task. These are all good tools. Sometimes, however, lighting installations do not fit the information supplied in the charts. If that is the case, other calculations must be done. Use the point-by-point or zonal cavity methods.

Point-by-Point Method

The point-by-point method is suitable for nonuniform lighting. It can be used to calculate the illumination for any surface from any angle, but the surface must be defined in size and must not be too big. Surfaces are usually walls, tabletops, or objects. The point-by-point method calculates the footcandles falling on a surface. It is based upon the requirements that only one source be calculated at a time and that the light from the source must come to the surface directly, not reflected from somewhere else. To

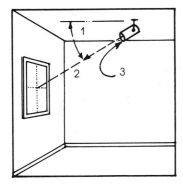

(1) Cosine of angle, (2) Distance,
(3) Candlepower

perform the calculations, three pieces of information must be known—the candlepower of the light source, the cosine of the aiming angle of the light, and the distance from the light source to the surface.

Point-by-point Formula

$$\text{footcandles} = \frac{\text{candlepower} \times \text{cosine of aiming angle}}{\text{distance}^2}$$

For instance, a picture 40 in. (102 cm) high and 30 in. (76 cm) wide, hung on the wall centered 3 ft (0.9 m) down from ceiling needs to be lighted. The light source is a ceiling-mounted adjustable fixture with a 75-watt PAR spot bulb, 3 ft (0.9 m) from the wall aimed at a 45° angle.

Calculate the footcandles at the center of the picture with the formula. The candlepower of the bulb is determined from the manufacturers' charts and is 4,500. The cosine of the aiming angle is determined from mathematics tables of cosines and is 0.7. The distance from the source to the picture is the square root of the sum of the square of the other two sides (3 ft squared + 3 ft squared) and is 18. This number can be determined by the Pythagorean theorem because the triangle is a right-angle triangle and the two sides are equal—three feet out from the wall (one side of the triangle) and three feet down (the other side of the triangle).

The calculations would be:

$$\text{footcandles} = \frac{4,500 \times \text{cosine of } 45°, \text{ or } 0.7}{18} = \frac{3,150}{18} = 175$$

For example, if the surrounding light was 55 footcandles or less, the picture illuminated to 175 footcandles would attract attention, because the rule of thumb for attracting attention is that the lighted object must be three times as bright as its surroundings. Therefore, in a residence a 75-watt PAR might be too bright and a smaller wattage could be used. On the other hand, if the painting was hanging in a shop window and the next shop was illuminated at 250 footcandles, the 75-watt bulb would not put out sufficient light to attract attention.

Zonal Cavity Method ·

Zonal cavity calculations are suitable for uniform lighting and a whole space can be lighted with this method. This method can determine the number of light fixtures needed to provide a specified footcandle level uniformly. It can also be rearranged to determine how many footcandles will be received from a certain number of luminaires with a certain light-producing capability. As part of the calculation procedure, a room is divided into cavities. Minimally, there is always a room cavity. For instance, a room with recessed ceiling luminaires and a specific footcandle level on the floor has a cavity of the whole room. At the other extreme, a room with suspended luminaires and tabletop work surfaces with a specific footcandle level would have three cavities—ceiling cavity (from the lower edge of the luminaires to the ceiling), room cavity (from the lower edge of the luminaires to the tabletops), and floor cavity (from the tabletop to the floor). Calculations would be made for each cavity.

The zonal cavity method relies on ratios of room size (length, width), height of the cavity,

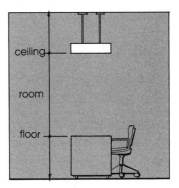

Zones for zonal cavity calculations.

Room cavity zone only.

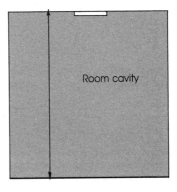

room reflectances, and depreciation of the light over time (light loss). In this calculation method, light is expected to be reflected from the surfaces in the room as well as to be direct. Eight pieces of information must be known—the type of luminaire, the footcandle level, the room size, the room reflectances, the height of the cavity, a chosen light source, the amount of lumens from that source, and if possible, light loss factors.

Light loss is caused by dirt in the room and the aging of the light bulb. Dirt in the room not only dulls reflectance possibilities of the major interior surfaces but also accumulates on luminaires. The design of luminaires makes some more susceptible to dirt deposits than others. The *IES Lighting Handbook 1981 Reference Volume* or *Student Reference* contains information on light loss factor. Further, light bulbs have different abilities to continue producing their lumens. Light bulb manufacturers publish such information as lamp lumen depreciation amounts.

With the zonal cavity method, determine the amount of footcandles needed by the amount of light required to do the task. Determine the room sizes and room reflectances. Determine the coefficient of utilization. Tables of coefficients are published by manufacturers. To use these tables, room size must be translated into a cavity ratio. The ratio is used to choose an appropriate coefficient with the appropriate room reflectance.

$$\text{cavity ratio} = \frac{\text{height of cavity} \times (\text{length of room} + \text{width of room})}{\text{length} \times \text{width}}$$

% Effective Ceiling Cavity Reflectance	80			50			10		
% Wall Reflectance	50	30	10	50	30	10	50	30	10
Room Cavity Ratio	Coefficients of Utilization								
1	.60	.57	.55	.53	.51	.50	.46	.45	.44
2	.53	.49	.46	.47	.44	.42	.41	.39	.38
3	.47	.42	.39	.42	.39	.36	.37	.35	.33
4	.42	.37	.34	.38	.34	.31	.34	.31	.29
5	.37	.32	.29	.34	.30	.27	.30	.27	.24
6	.34	.29	.25	.31	.27	.24	.28	.25	.22
7	.31	.26	.22	.28	.24	.21	.25	.22	.20
8	.28	.23	.20	.25	.21	.19	.23	.20	.17
9	.25	.20	.17	.23	.19	.16	.21	.18	.15
10	.23	.18	.15	.21	.17	.15	.19	.16	.14

E.T.L. REPORT NO. 379117

Testing laboratory's coefficient of utilization table. Reprinted by permission of Lightolier.

Determine the lumens per light bulb and how many bulbs are to be used in each luminaire. Lumens are published by manufacturers and are available in technical catalogs.

Zonal Cavity Formula

$$\frac{\text{number of fixtures}}{} = \frac{\text{footcandles desired} \times \text{total room area}}{\text{light bulbs per luminaire} \times \text{lumens per light bulb} \times \text{coefficient of utilization} \times \text{light loss}}$$

The zonal cavity method permits designers to light a space uniformly with specific footcandles on any horizontal surface, such as on a library tabletop or on the floor. Lighting is a science as well as an art. Often, the science of lighting is underused. It is not difficult to master at least one calculation method. The effects of predetermining and designing lighting at specific levels are powerful. Do not ignore them. Lighting actually can be more artful when scientifically designed.

bibliography

Butler, H., ed. *Home Decorating Using Light.* London: Marshall Cavendish, 1974.

Durrant, D. W., ed. *Interior Lighting Design.* London: Lighting Industry and the Electricity Council, 1973.

Electricity Council. *Better Office Light.* EC 2873R. London, 1972.

Evans, Nancy. "Lighting for Safety and Security." *Light Magazine,* 39, no. 2 (1970).

Fahsbender, Myrtle. *Residential Lighting.* New York: Van Nostrand, 1947.

Florence, Noel. "The Energy Effectiveness of Task-Oriented Office Lighting Systems." *Lighting Design and Application,* 9, no. 1 (January 1979).

Flynn, John F. "A Study of Subjective Responses to Low Energy and Nonuniform Lighting Systems." *Lighting Design and Application,* 7, no. 2 (February 1977): 6–15.

Flynn, John F. & Samuel M. Mills. *Architectural Lighting Graphics.* New York: Reinhold, 1962.

Freeth, Richard. *Plan Your Home Lighting.* London: Studio Vista, 1970.

General Electric Company. *Industrial Lighting.* Technical Paper 108R. Cleveland, 1977.

General Electric Company. *The Light Book.* Technical Paper. Cleveland, 1982.

General Electric Company. *Office Lighting.* Technical Paper 114R. Cleveland, 1977.

General Electric Company. *Store Lighting.* Technical Paper 106. Cleveland, 1970.

Gilliatt, Mary, and Douglas Baker. *Lighting Your Home: A Practical Guide.* New York: Pantheon, 1979.

Helms, Ronald N. *Illuminaton Engineering for Energy Efficient Luminous Environments.* Englewood Cliffs: Prentice-Hall, 1980.

Hopkinson, Ralph G., and John D. Kay. *The Lighting of Buildings.* London: Faber and Faber, 1972.

Illuminating Engineering Society. "Design Criteria for Lighting Interior Living Spaces." *IES Lighting Handbook, 1981 Application Volume.* New York: IES, 1981.

Illuminating Engineering Society. *IES Lighting Handbook, 1981 Reference Volume and Student Reference.* New York: IES, 1981.

Illuminating Engineering Society, Daylight Committee. "Recommended Practice of Daylighting." *Lighting Design and Application*, 9, no. 2. (February 1979).

Jones, Gerre. *How to Market Professional Design Services*. New York: McGraw-Hill, 1973.

Kalff, Louis C. *Creative Light*. New York: Van Nostrand, 1970.

Kamm, Lloyd J. *Lighting to Stimulate People*. Boston: Christopher Publishing House, 1948.

Lam, William. *Perception and Lighting as Form-givers to Architecture*. New York: McGraw-Hill, 1977.

Lightolier Incorporated. *Notes on the Lighting of Walls, Pictures, Draperies, and Other Vertical Surfaces*. Application Guide 10.02. Jersey City: 1974.

National Lighting Bureau. *Lighting Energy Management in Retailing*. Washington, DC: 1982.

Nuckolls, James. *Interior Lighting for Environmental Designers*. New York: John Wiley, 1976.

Pelger, Martin. *The Dictionary of Interior Design*. New York: Bonanza, 1966.

Phillips, Derek. *Planning Your Lighting*. London: Design Council, 1976.

Wells, Stanley. *Period Lighting*. London: Pelham, 1975.

Williams, H. G. "Office Lighting and Energy." *Skyscraper Management*. Washington DC: Building Owners and Managers Association Institute (April 1975).

White, Edward T., and Arthur I. Rubin. *Tracing Lighting Design Decisions for Open Office Space: A Pilot Study*. Washington DC: National Bureau of Standards, 1981.

Zackrison, Harry B., Jr. "Outside Lighting Systems Design." *Lighting Design and Application*, 10, no. 5 (May 1980).

index